Acta Neurochirurgica
Supplements

Editor: H.-J. Steiger

Neurosurgical Re-Engineering of the Damaged Brain and Spinal Cord

Edited by
Y. Katayama

Acta Neurochirurgica
Supplement 87

SpringerWienNewYork

Prof. Y. Katayama
Department of Neurological Surgery,
Nihon University, School of Medicine, Tokyo, Japan

Typesetting: Asco Typesetters, Hong Kong
Printing: A. Holzhausen, A-1140 Wien
Binding: Fa. Papyrus, A-1100 Wien

Printed on acid-free and chlorine-free bleached paper

SPIN: 10908266

CIP data applied for

With partly coloured Figures

ISSN 0065-1419
ISBN 3-211-00920-5 Springer-Verlag Wien New York

Preface

This volume is the second in a new series of proceedings covering the official scientific meetings of the Neurorehabilitation Committee of the World Federation of Neurosurgical Societies (WFNS). The first scientific meeting of the WFNS Neurorehabilitation Committee was held successfully in Munster, Germany, in 2000 under the auspices of Professor Klaus R. H. von Wild. The proceedings of that meeting (Functional Rehabilitation in Neurosurgery and Neurotraumatology) were published as a supplement to Acta Neurochirurgica (volume 79, 2001). This first scientific meeting highlighted the important role played by neurosurgeons in neurorehabilitation beginning at an early period after brain or spinal cord damage. Due to the great success of the first meeting and in expectation of rapid progress in the relevant fields of neurosurgery, the delegates of the WFNS Neurorehabilitation Committee decided to organize a further meeting two years later. The second scientific meeting, on which this volume is based, was thus held in Tokyo, Japan, on July 10–12, 2002.

Recent advances in functional neurosurgery have opened up an important new field in which neurosurgeons can collaborate fruitfully with specialists in the multi-disciplinary aspects of neurorehabilitation. Such collaboration involves a wide variety of neurosurgical techniques, including selective interruption of various neural circuits, stimulation of the cerebral cortex, deep brain structures, spinal cord and peripheral nerves with implantable stimulation systems, direct application of various drugs to the brain and spinal cord with implantable drug delivery systems, and cell transplantation as well as nerve grafting. These procedures provide benefits for the control of a wide variety of disabilities, such as those caused by movement disorders, motor palsy, spasticity, epilepsy, central pain, bladder dysfunction, speech and hearing disturbances, and the minimally conscious state. For these reasons, neurosurgeons have become increasingly involved in neurorehabilitation in recent years. The task carried out through the collaboration of neurosurgeons and specialists in neurorehabilitation cannot be viewed simply as a restoration of function or reconstruction of structure. Recent advances in neuroimaging techniques have begun to demonstrate that it involves extensive functional and structural reorganization of neural networks within the brain and probably the spinal cord. On this basis, we felt that it might be more appropriate to refer to such activities as re-engineering of the damaged brain and spinal cord. In order to encapsulate such a concept, the second scientific meeting was entitled the Second International Symposium on Neurosurgical Re-engineering of the Damaged Brain and Spinal Cord (NRDBS'02).

As a pre-symposium activity of the NRDBS'02, participants visited the Chiba Ryogo Center, which is one of the most well equipped institutes specializing in neurorehabilitation in Japan. We were there offered the opportunity to see and examine patients with severe disability caused by brain damage, and to discuss the neurosurgical techniques which might be potentially useful for each of the patients to improve their daily activities. During this visit to the Chiba Ryogo Center, we came to recognize that there are numerous neurosurgical procedures which would seem to be useful for improving the daily activities of patients with brain damage, but which tend to escape the physician's and even neurosurgeon's notice in everyday practice. The value of such neurosurgical techniques can be recognized and realized only by the involvement of neurosurgeons who are specialized in this field of neurorehabilitation at the point of clinical practice.

In order to facilitate further communication and cooperation between neurosurgeons and specialists in neurorehabilitation, the NRDBS'02 was held as a joint meeting with the Japan Coma Society, and the participation of a wide range of specialists in neurorehabilitation at the NRDBS'02 was encouraged. Although some of the topics presented were not directly related to surgical techniques, these topics aroused considerable attention and interest among neurosurgeons regarding the importance of their role in neurorehabilitation. It is worthy of particular mention that neurosurgeons were almost always included as a member of the team in the newly developed multidisciplinary neurorehabilitation programs which were introduced at the NRDBS'02. Encouragement was also given to the utilization of video materials in the actual presentations, so that participants could discuss the roles of neurosurgical techniques on the basis of a clearer understanding of the problems and solutions. This helped both neurosurgeons and specialists in neurorehabilitation to recognize the importance of the neurosurgeon's skill and knowledge. As a result, I believe that the NRDBS'02 was successful in identifying and defining more clearly the role of neurosurgeons within the broad multi-disciplinary spectrum of neurorehabilitation, and in reviewing recent advances in a variety of neurosurgical techniques that can contribute to the improvement of the patient's quality of life.

I extend my thanks to all participants at the NRDBS'02 and to the authors who have contributed to this volume. The secretarial work of Doctor Takamitsu Yamamoto, Department of Neurological Surgery and Applied System Neurosciences, Nihon University School of Medicine, was indispensable in organizing the NRDBS'02. I acknowledge Doctor Morimasa Kono, Director of the Chiba Ryogo Center and Doctor Nobuo Oka, Chief Neurosurgeon of the Chiba Ryogo Center, for providing us with the opportunity to see their patients and to engage in fruitful and enjoyable discussions with their staff during the pre-symposium visit. I also thank Professor Klaus R. H. von Wild, Chairman of the WFNS Neurorehabilitation Committee, Department of Neurosurgery and Unit for Early Neurorehabilitation, Clemenshospital, Munster, Germany, and Professor Tetsuo Kanno, Secretary of the Japan Coma Society, Department of Neurosurgery, Fujita Health University, Aichi, Japan, for their invaluable advice and support for the NRDBS'02.

Yoichi Katayama, MD, PhD
Chairman of the Second Scientific Meeting (NRDBS'02)
WFNS Neurorehabilitation Committee
President of the Japan Coma Society

Contents

Listed in Current Contents

Prolonged coma, minimally conscious state and persistent vegetative state

Acta Neurochir (2003) [Suppl] 87: 3–6

Control of sleep and wakefulness by brainstem monoaminergic and cholinergic neurons

Y. Kayama and **Y. Koyama**

Department of Physiology, Fukushima Medical University School of Medicine, Fukushima, Japan

Summary

Noradrenergic projection originating in the locus coeruleus, serotonergic projection from the dorsal raphe nucleus, and cholinergic projection from neurons gathering in the laterodorsal tegmental nucleus and scattering in the pedunculopontine tegmental nucleus constitute three diffuse projection systems arising from the brainstem and innervating wide areas of the brain. They may function as controllers of sleep and wakefulness. We have investigated functional roles of the projections by recording neuronal activity in these brainstem nuclei, and by observing effects of stimulation of the brainstem nuclei. The projection from the locus coeruleus is an arousal system, since the noradrenergic neurons are active specifically during waking, and activation of the noradrenergic projection excites upper brain structures. Functions of the serotonergic projection are still mysterious, since its action on upper brain is inhibitory in spite of waking-specific activity of the neurons. A group of cholinergic neurons constitute a system to induce and maintain paradoxical sleep. The cholinergic projection may have another role, i.e. to induce a rapid, transient elevation of vigilance level by their phasic response to novel, unfamiliar stimuli.

Keywords: Sleep; wakefulness; monoaminergic projections; cholinergic projection.

Introduction

When Moruzzi and Magoun [11] proposed the theory of ascending reticular activating system, it was not known which neurons, or neuronal groups, were the real substrate of the activating system. Thereafter, anatomical studies have revealed three projection systems which can be candidates of the activating system, as they originate in the brainstem and innervate wide areas of the brain directly with branching axons. They are (1) noradrenergic projection originating in the locus coeruleus, (2) serotonergic projection from the dorsal raphe nucleus, and (3) cholinergic projection from neurons gathering in the laterodorsal tegmental nucleus and scattering in the pedunculopontine tegmental nucleus. We have investigated the functional roles of these three projection systems by recording neuronal activity in these brainstem nuclei [6, 10], and by observing effects of stimulation of the brainstem nuclei [5, 7, 8].

Identification of neurons

We used rats for experiments, since in this species noradrenergic and cholinergic neurons gather in respective small nuclei (locus coeruleus and laterodorsal tegmental nucleus) and are well segregated from each other. In contrast, in the cat brainstem, noradrenergic neurons do not form a clear locus coeruleus, but are scattered in the parabrachial area, where they are intermingled with cholinergic neurons. Thus, distinguishing noradrenergic and cholinergic mechanisms is very difficult in cats in either recording or stimulating experiments.

We recorded single neuronal activity in undrugged, head-restrained rats [6, 10]. In the brainstem two kinds of action potentials, that is, brief spikes similar to those recorded usually in the cerebral cortex and broad spikes characteristically recorded in the brainstem, were clearly distinguished. It had been suggested that the broad spikes recorded in the locus coeruleus or raphe nuclei were generated by noradrenergic or serotonergic neurons, respectively [1]. In the same way, after double staining of sections for a marker of neurons recorded from (biocytin) and a marker of cholinergic neurons (NADPH-diaphorase), we proved that the broad spikes in the laterodorsal tegmental nucleus were generated by cholinergic neurons [9]. Thus, we could identify the nature of encountered neurons through spike shape and its position in histological sections.

induced by laterodorsal tegmental nucleus stimulation ended abruptly after the cessation of stimulation. Effect of dorsal raphe stimulation was opposite to that of locus coeruleus; dorsal raphe induced a long-lasting inhibition (Fig. 2B) [7].

These effects were mimicked by iontophoresis of the transmitters (Fig. 2D); acetylcholine increased the firing rate with quick onset and offset, noradrenaline also induced an increase of firing, but in this case with gradual onset and offset. Serotonin induced a long-lasting suppression of firing.

Concluding remarks

Summarizing all these studies, we hypothesize regarding functions of the projections from the brainstem that the noradrenergic projection from the locus coeruleus is an arousal system, and it also elevates vigilance level transiently with phasic sensory responses, while function of the serotonergic projection is still mysterious, since these neurons are active specifically during waking but activation of this projection produces depression of the upper brain. One of the functions of the cholinergic projection is clearly that a group of cholinergic neurons constitutes a system to induce and maintain paradoxical sleep. The cholinergic projection may have another role to induce a rapid, transient elevation of vigilance level by their phasic response to novel, unfamiliar stimuli.

As described above, sleep and wakefulness, or vigilance level, may be controlled by these diffuse projection systems arising in the brainstem. The noradrenergic projection, and also cholinergic projection, may be substrates of the classical activating system; among them it seems that the noradrenergic projection may function as a rather tonic activating system, and the cholinergic projection as a rather phasic activating system, though the latter has another function, for example to induce and maintain paradoxical sleep.

References

1. Aghajanian GK, Vandermaelen CP (1982) Intracellular identification of central noradrenergic and serotonergic neurons by a new double labeling procedure. J Neurosci 2: 1786–1792
2. Aston-Jones G, Chiang C, Alexinsky T (1991) Discharge of noradrenergic locus coeruleus neurons in behaving rats and monkeys suggests a role in vigilance. Prog Brain Res 88: 501–520
3. El Mansari M, Sakai K, Jouvet M (1989) Unitary characteristics of presumptive cholinergic tegmental neurons during the sleep-waking cycle in freely moving cats. Exp Brain Res 76: 519–529
4. Jacobs BL, Azmitia EC (1992) Structure and function of the brain serotonin system. Physiol Rev 72: 165–229
5. Kayama Y, Negi T, Sugitani M, Iwama K (1982) Effects of locus coeruleus stimulation on neuronal activities of dorsal lateral geniculate nucleus and perigeniculate reticular nucleus of the rat. Neuroscience 7: 655–666
6. Kayama Y, Ohta M, Jodo E (1992) Firing of 'possibly' cholinergic neurons in the rat laterodorsal tegmental nucleus during sleep and wakefulness. Brain Res 569: 210–220
7. Kayama Y, Shimada S, Hishikawa Y, Ogawa T (1989) Effects of stimulating the dorsal raphe nucleus of the rat on neuronal activity in the dorsal lateral geniculate nucleus. Brain Res 489: 1–11
8. Kayama Y, Takagi M, Ogawa T (1986) Cholinergic influence of the laterodorsal tegmental nucleus on neuronal activity in the rat lateral geniculate nucleus. J Neurophysiol 56: 1297–1309
9. Koyama Y, Honda T, Kusakabe M, Kayama Y, Sugiura Y (1998) In vivo electrophysiological distinction of histologically identified cholinergic neurons by extracellular recording and labeling in rat laterodorsal tegmental nucleus. Neuroscience 83: 1105–1112
10. Koyama Y, Jodo E, Kayama Y (1994) Sensory responsiveness of "broad-spike" neurons in the laterodorsal tegmental nucleus, locus coeruleus and dorsal raphe of awake rats: Implications for cholinergic and monoaminergic neuron-specific responses. Neuroscience 63: 1021–1031
11. Moruzzi G, Magoun HW (1949) Brain stem reticular formation and activation of the EEG. Electroencephalogr Clin Neurophysiol 1: 155–173
12. Shosaku A, Kayama Y, Sumitomo I, Sugitani M, Iwama K (1989) Analysis of recurrent inhibitory circuit in rat thalamus: Neurophysiology of the thalamic reticular nucleus. Prog Neurobiol 32: 77–102
13. Steriade M, Datta S, Paré D, Oakson G, Curró Dossi R (1990) Neuronal activities in brainstem cholinergic nuclei related to tonic activation processes in thalamocortical systems. J Neurosci 10: 2541–2559

Correspondence: Yukihiko Kayama, Department of Physiology, Fukushima Medical University School of Medicine, 1 Hikari-ga-oka, Fukushima 960-1295, Japan. E-mail: y-kayama@fmu.ac.jp

Acta Neurochir (2003) [Suppl] 87: 7–10

Electrical treatment of coma via the median nerve

E. B. Cooper[1] and **J. B. Cooper**[2]

[1] Department of Neurological Surgery, University of Virginia Health System, Charlottesville, Virginia, USA
[2] Department of Neurology, Virginia Commonwealth University, Richmond, Virginia, USA

Summary

The median nerve is a portal to interact with the injured comatose brain. Peripheral nerve electrical stimulation has a central nervous system effect. Two to three weeks of right median nerve stimulation (RMNS) can hasten awakening from deep coma by increasing the dopamine levels. Three cases of electrically treated GCS-4 teenagers with acute diffuse brain injuries from motor vehicle crashes are presented by video. Pilot studies of RMNS for acute post-traumatic coma states have been done over the last ten years at East Carolina University and the University of Virginia. The neurophysiological effects of RMNS have been well documented at several neurosurgical centers in Japan using neuroimaging and spinal fluid assays. RMNS is a safe, inexpensive, non-invasive therapy for neuroresuscitation of coma patients. When employed early in the coma, the time in the ICU may be shortened and the quality of the final outcome may be enhanced.

Keywords: Coma; median nerve; electrical stimulation.

Introduction

The median nerve serves as a peripheral gateway to the central nervous system. The sensory distribution of the hand exhibits disproportionately large cortical representation [23, 28]. Within the brainstem, the ascending reticular activating system (ARAS) maintains wakefulness. The spinoreticular component of the median nerve synapses with neurons of the ARAS. Recovery from coma is driven by the electrically induced elevation of dopamine [14, 19, 26]. The right median nerve was chosen as a portal to stimulate the brainstem and cerebrum because increased alertness and better speech have been observed after RMNS [5, 25]. Broca's motor/speech planning area in the left frontotemporal region has been shown in positron emission tomography (PET) to become more active when a subject moves, or even contemplates moving his/her hand [18]. This process is mimicked in RMNS [25].

My interest in stimulation began 30 years ago with a study involving a paraplegic individual at University of Virginia (UVa) in Charlottesville, Virginia. Radio-linked, implanted electrodes strengthened muscles and produced crude ambulation [2]. From 1987 to 1989 individuals with quadriplegia were helped to use their forearm muscles through voice-activated electrical stimulation. This technique resulted in semi-voluntary hand opening and closing at the Department of Biomedical Engineering at Duke University in Durham, North Carolina [3, 7]. Significant improvement was noted in distal motor abilities in response to electrical stimulation [1, 9]. Proximal voluntary and contralateral arm increases in performance were also noted during strength testing [3].

Similar computerized electrical stimulation was applied to individuals with severe mental/motor delays at Caswell Center in Kinston, North Carolina, in hopes of improving function and awareness [10]. Progressive augmentation of mental awareness was noted. The observed cross-over effect in the quadriplegic population, along with the central arousal of the mentally challenged population, led to the postulation that stimulation of the median nerve causes significant central nervous system activation [5, 11–14].

Materials and methods

The electrical stimulators at the University of Virginia project and East Carolina University projects were Empi battery-powered units. They supplied trains of asymmetric biphasic pulses at an amplitude of 15–20 mA with a pulse width of 300 microseconds at 40 Hz for 20 sec/min. The treatment was done for 8–12 hours daily and 8 hours per day for 2–3 weeks.

The treatment usually started 1–2 weeks after the closed head injury with Glasgow Coma Scale (GCS) [16, 24] less than 8. The trains of pulses were delivered to the volar aspect of the right distal forearm over the median nerve via lubricated surface rubber electrodes measuring 2.5 by 2.5 centimeters. The pair of electrodes were embedded 2 cm apart in the midline of plastic cuffs from Carolina Ortho Prosthetics, Greenville, North Carolina.

For children and small adults, a setting of up to 15 milliamps is recommended and also for those adults with agitation at the 20 milliamp setting. For safety purposes, a plastic cap or clear tape should be applied over the external power dial on the electrical stimulator to prevent inadvertent over stimulation [6].

The right median nerve motor stimulation was approximately 1.5 times the motor threshold. This usually produced strong right thenar abduction and flexion of the index and middle fingers plus some wrist flexion. Temporary mild elevation of the vital signs is not unusual, but intracranial pressures usually remain stable.

In the first few days of treatment, mirror movements of the unstimulated left hand may occur. This dynamic cross-over effect heralds reactivation of the cerebral hemispheres through the corpus callosum in the electrically stimulated comatose patient. Usually the first simple command that the patient will respond to after one or two weeks of RMNS is a sluggish opposition of the right thumb and index finger.

This purposeful hand response while the brain injury patient still appears to be semi-comatose, demonstrates that the 5 million electrical pulses delivered to the nervous system in the first ten days of treatment have been copied and stored in the hard drive of the brain.

Results

Over the past ten years (during the "Decade of the Brain"), a pilot project at East Carolina University (ECU) [5] and two pilot studies at the University of Virginia (UVa) [5, 20] have been reported. In two of the three studies, earlier awakening in the electrically treated comatose patients has been noted compared to control or historical patients. By several measures: days in coma, days in the intensive care unit, and days of endotracheal intubation, have been shorter in the treated group [5, 20].

Three young decerebrate comatose patients (GCS-4) will be presented by video. These cases are presented to show the possibility of functional recovery with RMNS in situations where the initial prognosis was very grim.

1. CI, a 16 year old female, was involved in an MVC and sustained severe closed head injury in 1994. She suffered a basilar skull fracture, cerebrospinal fluid otorrhea, left facial fracture, and left pelvic fracture. CT scan revealed left internal capsule contusion, right cerebellar subarachnoid hemorrhage, and blood in the fourth ventricle. Decerebrate posturing was observed and she received a GCS of 4. She was briefly given electrical stimulation but intracranial pressures continued to rise. With her extremely poor prognosis, she was expected to die and was extubated. She breathed spontaneously and electrical stimulation was resumed to the right median nerve. Within 1 week of stimulation, she exhibited semi-purposeful movement of her right arm and leg and scored 7 on the GCS. After a total of 2 weeks of stimulation, she was out of coma. This increase of 6 points was consistent with the UVa pilot project observation [4]. One month after the injury, CI followed simple commands. At 2 months post-injury CI could walk with assistance and read aloud. Two years later CI talked and walked well. She resumed dancing and driving. She graduated from college with a "B" average and has been working as a recreation director for two years.

2. In March of 2000, a 12 year old boy (KF) was struck by a van and had severe brain, intra-abdominal, and multiple extremity fractures including compound fractures of the pelvis. On his initial CT scan there was left frontal contusion, a small amount of subarachnoid hemorrhage in the interpeduncular cistern and a non-depressed skull fracture of the left parietal bone. There was also a fracture of the left temporal bone. On the follow-up scan two days later, there were several contusions (right frontal and left temporal), increased edema with marked effacement of the cortical sulci, and intraventricular hemorrhage. On the scan one week post-injury there was diffuse brain swelling and multiple hemorrhagic shearing injuries and hemorrhagic contusions of the left frontal lobe extending into the temporal lobe. KF underwent several abdominal and orthopaedic operations. He remained comatose with elevated intracranial pressures (over 70 mm Hg) in spite of two courses of barbiturate therapy. Pupils remained unequal. Survival was very questionable. Surface electrical stimulation in the 15 milliamps range to the left median nerve was commenced two weeks post injury. The right forearm was in a cast. After two weeks of daily stimulation, he began to emerge from the coma. He progressively improved and regained his ability to speak. He could use his hands, in spite of a right hemiparesis he could use both hands. Two months post-injury, he was transferred to a rehabilitation center. Electrical stimulation was resumed, but switched to the right median nerve to help reduce the right hemiparesis. He continued to improve and was discharged home six weeks later. He started home schooling and made good grades. Last year he had a "B" average in junior high school and was able to resume some athletic activities.

3. AT, a 14 year old female, was involved in a motor vehicle collision and sustained a severe closed head injury in 1995. She had a hemothorax and a pulmonary

contusion. Her right mandibular condyle penetrated the base of her skull through the temporal bone into the middle cranial fossa. CT scan showed haemorrhagic foci in the left cerebral hemisphere. She exhibited alternating decerebrate and decorticate posturing and received a GCS of 4. Within the first week of stimulation she began gripping spontaneously. After 1 week of stimulation her GCS was 6. After 2 weeks of stimulation she began to open her eyes spontaneously and received a GCS of 8. At 2 months post-injury she was eating well and speaking. Within 5 months she was playing volleyball and doing well in school. She recently graduated from college with a "B" average.

Discussion

In the textbook by Wilkins and Rengachary, Neurosurgery, 1996, there is a chapter written by Marshall and Marshall on "outcome prediction in severe head injury" [17]. There were 49 GCS-4 survivors. Seven had a Good outcome (14%), 9 had Moderate Disability (18%), for a total of 32% favorable outcome.

There were 12 GCS-4 survivors treated with RMNS at ECU. They were followed for at least one year. A satisfactory result (Good in 4 patients and Moderate recovery in 3) was reached by one year in 58% of the GCS-4 patients treated with right median nerve electrical stimulation [25].

Through maintenance of existing neuronal circuitry, earlier awakening from coma may lead to a higher final level of function. Increased cerebral activity, as observed in RMNS, may also facilitate synaptogenesis in damaged cerebral cortex [8, 21]. The clinical observations indicate that RMNS has a beneficial effect on the resumption of language capabilities, possibly through stimulation of Broca's motor speech area [22, 25].

The patient may obtain a better outcome as a result of the stimulation [5, 19, 22]. The observations gained from the treated series of patients at ECU, Fujita Health University, Nihon University and the pilot studies at UVa suggest that RMNS may have a positive effect on brain-injured comatose individuals. Non-invasive median nerve electrical stimulation is easily employed with little risk and is cost effective [4, 27]. This new technique can improve neurological outcomes.

References

1. Bertoti D (2000) Electrical Stimulation: A reflection on current clinical practices. Assist Technol J 12(1): 21–32
2. Cooper E, Bunch W, Campa J (1973) Effects of chronic human neuromuscular stimulation. Surg Forum 24: 477–479
3. Cooper E, Han D, McElhaney J (1988) A voice controlled computer system for restoring limited hand functions in quadriplegics. Proceedings of the American Input Output Systems Applications Conference, San Francisco
4. Cooper E, Cooper J, Jane J (1996) Right median nerve electrical stimulation of comatose patients, unpublished article, presented at 5th Annual Meeting for the Society for Treatment of Coma, Sendai, Japan
5. Cooper J, Jane J, Alves W, Cooper E (1999) Right median nerve electrical stimulation to hasten awakening from coma. Brain Injury 13: 261–267
6. Focus Instruction Manual (1999) Empi, Inc, St Paul
7. Gersh M (1992) Electrotherapy in rehabilitation. FA Davis Co, Philadelphia
8. Grady M, Jane J, Steward O (1989) Synaptic reorganization within the human central nervous system following injury. J Neurosurgery 71: 534–537
9. Hamilton G, Cooper E (1987) Functional electrical stimulation (FES) in paralytic hand musculature. Proceedings of the Tenth International World Confederation of Physical Therapy, Sydney, Australia
10. Han D (1992) The voice operated electrical stimulation system. The Proceedings of the Johns Hopkins National Search for Computing Applications to Assist Persons with Disabilities, Baltimore
11. Hirata Y, Ushio U (2000) A case of successful treatment by median nerve stimulation for prolonged moderate consciousness disturbance in the chronic phase following subarachnoid hemorrhage. Proceedings of the Society for Treatment of Coma 9: 85–89
12. Kanno T *et al* (1989) Effects of dorsal spinal cord stimulation (DCS) on reversibility of neuronal function–experience of treatment for vegetative states. Pace 12: 733–738
13. Kanno T (August 1996) personal communication
14. Kanno T (June 2001) personal communication
15. Levin H, Jane J (1992) Severe head injury in children: experience of the traumatic coma data bank. Neurosurgery Sept (3)31: 435–443
16. Liebert MA (2000) Glasgow coma scale score. J Neurotrauma (6/7) 563–571
17. Marshall F, Marshall S (1996) Outcome prediction in severe head injury, chap 267 in neurosurgery. In: Wilkins R, Rengachary S (eds) McGraw-Hill, New York
18. Montgomery G (1989) The mind in motion. Discover 10: 58–64
19. Moriya T, Hayashi N *et al* (1999) Median nerve stimulation method for severe brain damage, with its clinical improvement. Proceedings of the Society for Treatment of Coma 8: 111–114
20. Peri C, Shaffrey M, Farace E *et al* (2001) Pilot study of electrical stimulation on median nerve in comatose severe brain injured patients: 3-month outcome. Brain Injury 15(10): 903–910
21. Robertson I, Murre J (1999) Rehabilitation of brain damage: brain plasticity and principles of guided recovery. Psychology Bull, Sept (5) 125: 544–575
22. Scherder E, Bouma A, Steen L (1992) Influence of transcutaneous electrical nerve stimulation on memory in patients with dementia of the Alzheimer type. J Clin Exp Neuropsychol 14: 951–960

23. Spiegel J, Treede RD *et al* (1999) Functional MRI of human primary somatosensory and motor cortex during median nerve stimulation. Clin Neurophysiol 110(1): 47–52
24. Teasdale G, Jennett B (1974) Assessment of coma and impaired consciousness, a practice scale. Lancet July 2: 81–84
25. University of Virginia and East Carolina University (1998) Coma stimulation projects. Unpublished data
26. Yamamoto K *et al* (1997) A case of persistent vegetative state treated with median nerve stimulation. Proceedings of the Society for Treatment of Coma 6: 117–121
27. Yokoyama T, Kamei T, Kanno T (1996) Right median nerve stimulation for comatose patients. Proceedings of the Society for Treatment of Coma 5: 117–125
28. Yousry T *et al* (1995) Topography of the cortical motor hand area: prospective study with functional MR imaging and direct motor mapping at surgery. Radiology 195(1): 23–29

Correspondence: Edwin B. Cooper, M.D., Clinical Associate Professor, Department of Physical Medicine and Rehabilitation, Brody School of Medicine, East Carolina University, Greenville, North Carolina, USA. E-mail: edwincooper@earthlink.net

Acta Neurochir (2003) [Suppl] 87: 11–14

Regaining consciousness for prolonged comatose patients with right median nerve stimulation

J. T. Liu[1], C. H. Wang[1], I. C. Chou[1], S. S. Sun[2], C. H. Koa[2], and E. Cooper[3]

[1] Department of Neurosurgery, China Medical College Hospital, Taichung, Taiwan R.O.C
[2] Department of Nucleomedicine, China Medical College Hospital, Taichung, Taiwan R.O.C
[3] Department of Physical Medicine and Rehabilitation, East Carolina University, Greenville, NC, USA

Summary

During the past half year, we have utilized right median nerve stimulation (produced by Empi. Co. USA) to awaken consciousness of patients on 6 cases, 2 with brain trauma, one with aneurysm rupture, one with hemorrhagic stroke and two with hypoxic encephalopathy. The comatose duration ranged from 6 to 17 weeks. All patients underwent SPECT scan for cerebral perfusion evaluation and neurotransmitter quantification before and after the stimulation. The stimulation time lasted from 8 to 10 hours on daytime.

The stimulation persisted for 3 months in all patients. Four patients recovered from the consciousness within 35 days. There was no obvious clinical improvement in two patients. Brain perfusion increased in all cases through the SPECT study after stimulation. The elevation of neurotransmitter in CSF was found in 5 out of the 6 cases.

Median nerve stimulation elevates the cerebral blood flow and may influence the patient's consciousness. Young patients (<40 years old) had better results than old patients. The stimulation could be initiated if the patient's acute stage was over. In our series, although increase in cerebral blood flow was found in all cases, some patients did not regain consciousness.

Keywords: Median nerve stimulation; prolonged comatose patient; cerebral blood flow SPECT scan.

Introduction

Dorsal column stimulation of the cervical spinal cord has been widely used to treat prolonged comatose patients [6, 7, 10]. The procedure can be performed through the percutaneous route or open surgery, however, it is an invasive procedure. The whole installation fee, including the stimulator, surgical procedure and hospitalization, costs about USD 20,000 in Taiwan.

The mechanism of right median nerve stimulation to awaken consciousness was first published by Dr. J. B. Cooper [2] in 1999, and since then, many papers have described this method [3, 6, 15]. Two electrodes were pasted on the palmar side of the right wrist joint to stimulate the median nerve. We performed right median nerve stimulation on 6 comatose patients. Cerebral blood perfusion increase was found in both sides of the cortex, basal ganglion, thalamus and hemisphere by SPECT method. The clinical result showed that 4 in 6 patients regained consciousness after stimulation.

Materials and methods

From October 1, 2001 to March 31, 2002, we stimulated 6 comatose patients with an electric stimulator (produced by Empi. Co. USA) on the right median nerve to awaken consciousness. There were two cases with brain trauma, one case with aneurysm rupture, one case with hemorrhagic stroke and two cases of hypoxic encephalopathy. These patients underwent standard treatments with or without surgical craniotomy. There were five males and one female. The age of the patients ranged from 1.25 to 66 year-old; mean age: 27.3 years (Table 1). The comatose score was between 5 and 9 before stimulation according to the Glasgow Coma scale. The duration of coma ranged from 6 to 17 weeks. No patients stayed in ICU and vital signs were stable. The brain CT (Computerized Tomography) scan of all patients showed mild to moderate brain atrophy.

A transcutaneous electric stimulator pasted on the palmar side of the right wrist skin supplied trains of asymmetric biphasic pulses at an amplitude of 20 mA with a pulse width of 300 μs at 35 Hz for 20 sec on/50 sec off. The stimulation time was 10 hours (7–11/am and 1–7/pm) on daytime per day during the comatose stage. The stimulation time was changed to 8 hours (7–11/am and 2–6/pm) when any patient became conscious. The stimulation persisted for 3 months in all patients.

Cerebral perfusion was evaluated with SPECT scan (HMPAO Tc-99m). Three separate cerebral perfusion studies were performed for each patient, before and after stimulation (one and three months). CSF was collected just one hour before SPECT scanning and Neurotransmitter (Dopamine) was subsequently quantified by the radioisotope method.

Table 1. *Patient population*

	Age	Sex	Injured category	Surgery	CT and SPECT scan	Comatose period
1	20 Y/O	M	trauma	craniotomy V-P shunting	mild brain atrophy, diffused decrease of blood flow	7 weeks
2	28 Y/O	M	trauma	craniotomy	moderate brain atrophy, diffused decrease of blood flow	17 weeks
3	66 Y/O	M	aneurysm rupture	craniotomy	mild brain atrophy, diffused decrease of blood flow	6 weeks
4	34 Y/O	M	hemorrhagic stroke	craniotomy	blood flow decrease in left hemisphere	10 weeks
5	15 M/O	F	hypoxia	V-P shunting	diffused brain atrophy and decreased perfusion	13 weeks
6	19 Y/O	M	hypoxia	V-P shunting	marked brain atrophy and decreased perfusion	12 weeks

Table 2. *Results of right median nerve stimulation*

	Consciousness recover time	SPECT scan	Dopamine in CSF	Clinical appearance after 3 months stimulation
1	30th day	increased	increased	Walk
2	28th day	increased	increased	Utter a few words, N-G feeding
3	23rd day	increased	increased	Walk, Utter a few words
4	35th day	increased	increased	Utter a few words
5	Coma	increased	increased	Sucking reflex crying appear
6	Coma	increased	unchanged	unchanged

Results

Clinical assessment was performed by a neurologist and a rehabilitation doctor (Table 2). The efficacy of treatment was recorded at regular intervals (pre- and post-stimulation, one, two and three months) by video and photography. EEGs were studied in two hypoxic encephalopathic patients because marked brain atrophy was found in CT scan. There was significant increased perfusion on both hemispheres in all candidates after stimulation.

Four patients gained the consciousness between day 23 and 35 after initial stimulation. All (case 1, 2, 3, 4) underwent aggressive physical training which included building muscle power, walking and language training. Two patients (case 1, 3) can now walk with the help by a walker. Two patients (case 2, 3) can utter some meaningful words. Three patients (case 1, 3, 4) take their meals without N-G tube feeding. One patient (case 2) who had been comatose for 17 weeks recovered consciousness on the 35th day after stimulation. After a series of training, the patient still can not stand up or feed normally.

Two hypoxic patients (case 5, 6) had no clinical improvement, although the cerebral perfusion increased after stimulation in both cases. The CT image showed moderate brain atrophy. In case 5, the child developed oral sucking ability and crying response on the 40th and 48th day respectively, although she was still comatose. Dramatic increase of cerebral perfusion was found after stimulation (Fig. 1). EEG showed more focal discharges than before stimulation (Fig. 2).

Dopamine elevation in CSF was found in five cases: one patient (case 6) had no change. Post-stimulation EEG presented no change in case 6.

Four patients (case 1, 3, 4, 5) developed skin irritation due to long term attachment to electric pad and one case (case 1) developed epilepsy on the 4th day after stimulation because serum anticonvulsant levels were lower than therapeutic levels. There was no seizure after anticonvulsant dose adjustment. No major or uncontrolled complications were found.

Discussion

Regaining consciousness with spinal cord stimulation in vegetative patients has been reported in many papers [6, 7, 10]. Nevertheless, it is not a popular procedure because of economical reasons and variant results. JB Cooper reported that right median nerve stimulation may help acutely brain-injured patients recover from coma earlier. The possible mechanism of reactivating the brain cell function includes a significant increase in cerebral blood flow and enhancement of neurotransmitter metabolism in brain cells [4, 8, 9, 11]. Brain cell metabolism can also be facilitated in damaged cortex [2].

In Japan, many papers have reported that median nerve stimulation can be used to facilitate the efficacy of physical therapy for central nervous system injured patients [3, 5, 13, 14]. Through SPECT scanning survey and neurotransmitter quantification in our patients results of median nerve stimulation were similar to those of spinal cord stimulation [6–8, 10]. The difference between Cooper's series and ours is that we

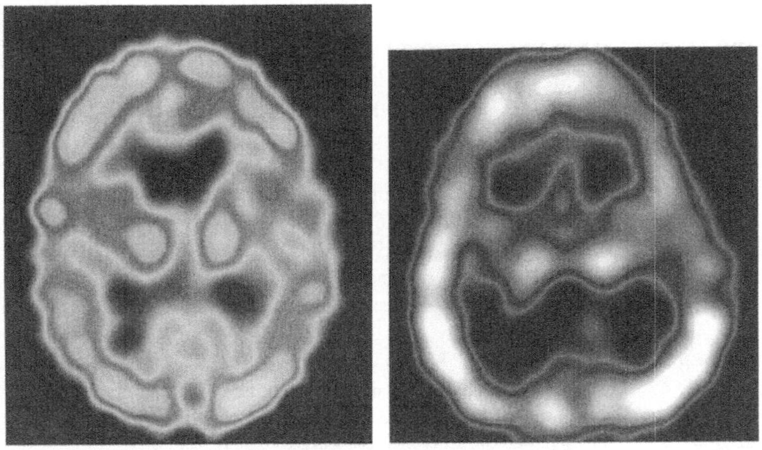

Before stimulation Three months after stimulation

Fig. 1. The cerebral perfusion increased after stimulation

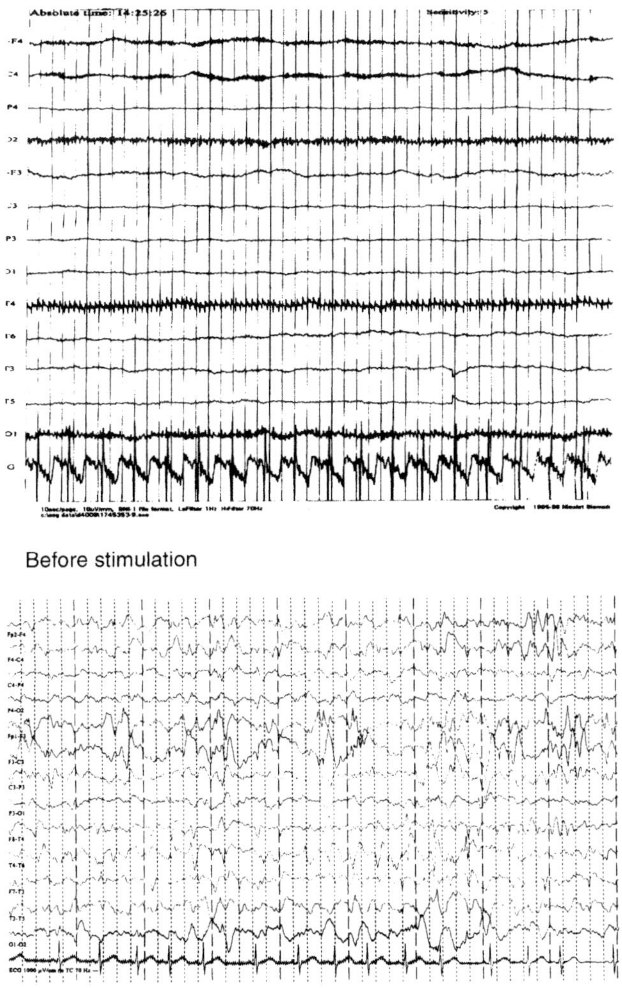

Before stimulation

Three months after stimulation

Fig. 2. More low voltage discharge were found after stimulation

treated patients in chronic stage rather than in acute stage [2]. However, there was not much clinical efficacy in the hypoxic encephalopathy group.

In our series, the procedure was still effective even if the patient was comatose during more than three months (17 weeks). It is different from some author's report [5, 12, 14, 15]. In case 5, the positive response after stimulation encouraged us that right median nerve stimulation might be effective in hypoxic encephalopathic group if the patient's age is younger than 6 years-old.

From the economic point of view, the cost is much less than that of spinal cord stimulation (USD 1,470 vs USD 20, 140). In addition, installation of the median nerve stimulator is not an aggressive procedure.

Conclusion

Median nerve stimulation increased the cerebral blood flow and might influence the patient's consciousness regardless of the comatose stage and period. Stimulation could be initiated if the patient is not in acute stage. In our series, although increased blood flow was found in all cases, some patients did not regain consciousness. For hypoxic encephalopathic patients, median nerve stimulation could be attempted to regain consciousness if the patient's age is younger than 6 years-old. This result encouraged the physician to determine an effective method of neurorehabilitation.

References

1. Bouma G, Muizelaar J (1993) Evaluation of regional cerebral blood flow in acute head injury by stable xenon-enhanced computerized tomography. Acta Neurochir (Wien) [Suppl] 59: 34–40
2. Cooper JB, Jane JA, Cooper EB *et al* (1999) Right median nerve electrical stimulation to hasten awakening from coma. Brain Injury 13: 261–267
3. Glanz M, Klawansky S, Stason W *et al* (1996) Functional electrostimulation in poststroke rehabilitation: a meta-analysis of the randomized controlled trials. Arch Phys Med Rehab 77: 549–553
4. Grady M, Jane J, Steward O (1989) Synaptic reorganization within the human central nervous system following injury. J Neurosurg 71: 534–537
5. Hirata Y, Ushio U (2000) A case of successful treatment by median nerve stimulation for prolonged moderate consciousness disturbance in the chronic phase following subarachnoid hemorrhage. The Society for the Treatment of Coma 9: 85–89
6. Hosobuchi Y (1985) Electrical stimulation of cervical spinal cord increases cerebral blood flow in humans. Appl Neurophysiol 48: 372–276
7. Kanno T, Kamel Y, Yokoyama T *et al* (1989) Effects of dorsal column stimulation (DCS) on reversibility of neuronal function – Experience of treatment for vegetative states. PACE 12: 733–738
8. Lawrence F, Eisenberg H, Jane J *et al* (1991) The outcome of severe closed head injury. Journal of Neurosurgery 75: s28–s36
9. Marion D, Darby J, Yonas H (1991) Acute regional cerebral blood flow changes caused by severe head injuries. J Neurosurg 74: 407–414
10. Meglio M, Cioni B, Visocchi F *et al* (1991) Spinal Cord Stimulation and Cerebral haemodynamics. Acta Neurochir (Wien) 111: 43–48
11. Obrist WD, Langfitt TW, Jaggi JL *et al* (1984) Cerebral blood flow and metabolism in comatose patients with acute head injury. J Neurosurg 61: 241–253
12. Peri CV, Shafery ME, Cooper ED *et al* (2001) Pilot study of electrical stimulation on median nerve in comatose severe brain injured patients: 3-month outcome. BRAIN INJURY 15: 903–910
13. Scherder E, Bouma A, Steen L (1992) Influence of transcutaneous electrical nerve stimulation on memory in patients with demential of the Alzheimer's type. Journal of Clinical and Experimental Neuropsychology 14: 951–960
14. Yamamoto K, Sugita S, Ishikawa K *et al* (1997) A case of persistent vegetative state treated with median nerve stimulation. The Society for the Treatment of Coma 5: 117–125
15. Yokoyama T, Kamei T, Kanno T (1996) Right median nerve stimulation for comatose patients. The Society for the Treatment of Coma 5: 117–125

Correspondence: Jung-Tung Liu, M.D., Neurosurgeon Department of Neurosurgury, China Medical College Hospital, No. 2, Yue-Der Road, 404, Area North, Taichung, Taiwan, R.O.C. E-mail: ljt7522h@ms34.hinet.net

Acta Neurochir (2003) [Suppl] 87: 15–18

DBS therapy for a persistent vegetative state: ten years follow-up results

T. Yamamoto, Y. Katayama, K. Kobayashi, M. Kasai, H. Oshima, and **C. Fukaya**

Department of Neurological Surgery and Division of Applied System Neuroscience, Nihon University School of Medicine, Tokyo, Japan

Summary

Twenty-one cases of a persistent vegetative state (PVS) caused by various kinds of brain damage were evaluated neurologically and electrophysiologically at 3 months after the brain injury. The 21 cases were treated by deep brain stimulation (DBS) therapy, and followed up for over 10 years. The stimulation sites were the mesencephalic reticular formation (2 cases) and CM-pf complex (19 cases). Eight of the patients emerged from the PVS, and became able to obey verbal commands. However, they remained in a bedridden state. These 8 cases revealed a desynchronization on continuous EEG frequency analysis. The Vth wave of the ABR and N20 of the SEP could be recorded even with a prolonged latency, and the pain-related P250 was recorded with an amplitude of over 7 μV. The mean survival time of these 8 cases was 6.1 years, as compared to 3.1 years for the other 13 cases. Overall, 4 cases are alive after more than 10 years. DBS therapy may be useful for allowing patients to emerge from a PVS, if the candidates are selected according to neurophysiological criteria. The fact that 19% (4/21) of the PVS cases treated with DBS survived for over 10 years should be stressed in comparison with the usual survival period for the untreated PVS.

Keywords: Vegetative state; deep brain stimulation; CM-pf complex.

Introduction

The Multi-Society Task Force on PVS (1994) summarized the medical aspects of the PVS [6, 7]. They made the statement that the vegetative state is a clinical condition of complete unawareness of the self and the environment, accompanied by sleep-wake cycles, with either complete or partial preservation of hypothalamic and brainstem autonomic function. In addition, patients in a vegetative state show no evidence of sustained, reproducible, purposeful, or voluntary behavioral responses to visual, auditory, tactile, or noxious stimuli; show no evidence of language comprehension or expression; have bowel and bladder incontinence; and have variably preserved cranial-nerve and spinal reflexes.

According to the above statement on the PVS, we selected 21 patients who remained in a persistent vegetative state (PVS) for more than 3 months after sustaining various kinds of brain damage. These 21 cases were evaluated neurologically and electrophysiologically at 3 months after the brain injury and were treated by deep brain stimulation (DBS) therapy [8, 9]. We have followed up these cases for over 10 years, and assess the long-term survival rate of PVS patients treated with DBS therapy.

Materials and methods

All of the 21 patients who were treated by chronic DBS therapy had been in a condition that satisfied the criteria of the PVS outlined by the Multi-Society Task Force on PVS (1994) for at least 3 months before undergoing DBS therapy, and could be followed up for more than 10 years after the DBS therapy. Their ages ranged from 19 to 75 years old. The causes of the initial coma were head injury (9 cases), cerebrovascular accident (9 cases) and anoxia (3 cases). At 3 months after the onset of the comatose state, neurological examinations and neurophysiological evaluations were carried out. The neurophysiological evaluations included assessments of the auditory brainstem response (ABR), somatosensory evoked potential (SEP), pain-related P250 and continuous EEG frequency analysis [5, 10].

The chronic DBS was applied using a chronically implanted flexible wire electrode inserted by stereotaxic surgery under local anesthesia. As target points for the chronic DBS, the mesencephalic reticular formation (2 cases) and the CM-pf complex (19 cases) were selected. The stimulation was given every 2 to 3 hours during the daytime, and was continued for 30 min at one time. The frequency of the stimulation was mostly fixed at 25 Hz, and the intensity was decided according to the responses of each patient, being at slightly higher than the threshold for inducing an arousal response. To apply the chronic DBS, we employed a chronically implanted flexible electrode (3380, Medtronic Co.) and a transmitter-receiver system (3470 and 3425, Medtronic Co.). The target point in the mesencephalic reticular formation was the nucleus cuneiformis, which is located in the dorsal part of the nucleus ruber and ventral part of the deep layer of the superior colliculus. The CM-pf complex was selected as the stimulating point in the non-specific thalamic nucleus (Fig. 1). We followed up the 21 cases for over 10 years, and assessed about the survival period. We also analyzed the effects of DBS treatment on PVS patients.

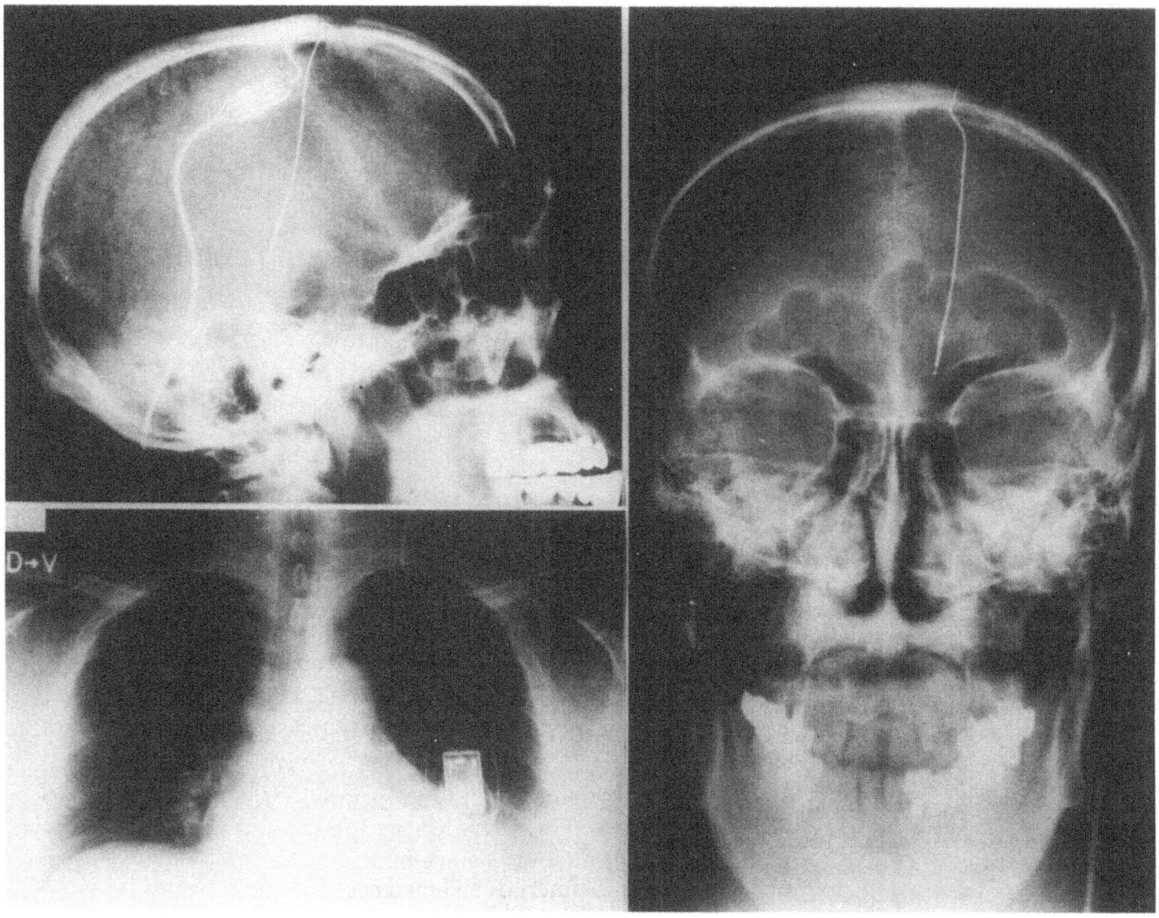

Fig. 1. Deep brain stimulation for the treatment of a vegetative state patient. The stimulating electrode was implanted for CM-pf stimulation

Results

Recovery from PVS

Eight of the 21 cases emerged from the PVS, and could communicate with some speech or other responses, but needed some assistance with their everyday life in bed. Even after long-term rehabilitation, their state of being bedridden remained unchanged. The other 13 cases were unable to communicate at all and failed to emerge from the PVS. In the 8 cases that emerged from the PVS following DBS therapy, the Vth wave of the ABR and N20 of the SEP were recorded even with a prolonged-latency; continuous EEG frequency analysis revealed a desynchronization pattern or slight desynchronization pattern; and the pain-related P250 was recorded with an amplitude of over 7 μV.

Survival period after DBS

A period of over 10 years follow-up has passed for all 21 DBS cases, and 4 cases survived for over 10 years. The survival rate among the 21 cases after the DBS therapy was 21 cases (100%) at 1 year, 16 cases (76%) at 2 years, 12 cases (57%) at 3 years, 10 cases (48%) at 4 years, 6 cases (29%) at 5 years, 5 cases (24%) at 6 years, 5 cases (24%) at 7 years, 5 cases (24%) at 8 years, 5 cases (24%) at 9 years, and 4 cases (19%) at 10 years. The difference in survival period between the group which recovered ($n = 8$) and the non-recovered group ($n = 13$) was significant ($p < 0.05$) at over ca. 6 years survival (Fig. 2). The causes of death were falling into infectious disease ($n = 15$), ileus ($n = 1$), and occurrence of cancer ($n = 1$).

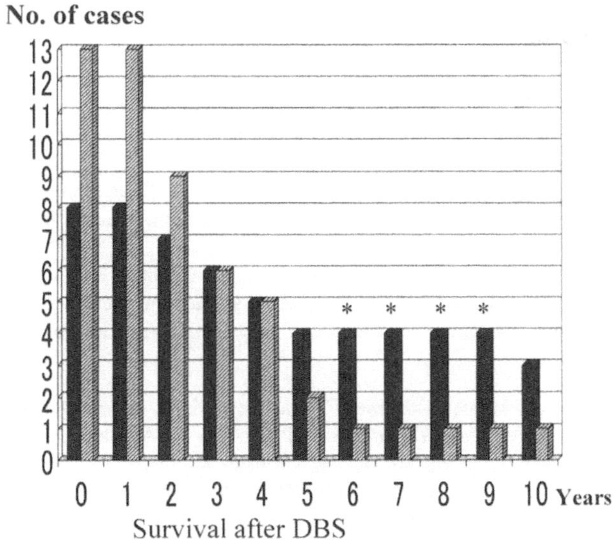

Fig. 2. Long-term follow-up results for the PVS after DBS therapy giving a comparison of the survival period between recovered and non-recovered cases. * $p < 0.05$ (Fisher's exact probability test). ■ Recovered; ▨ Non-recovered

Discussion

We were recently able to renew contact with the family of one patient who was unresponsive, and so to add 1 case to our long-term follow-up results for DBS therapy in the PVS [10]. DBS applied to the mesencephalic reticular formation or CM-pf complex can exert a strong arousal response and elicit marked increases in r-CBF and r-CMRO2. In the PVS, cerebrocortical functions are more disturbed than brainstem functions, and the relationship between the brainstem and cerebral cortex is important for maintaining consciousness. We therefore mainly selected the CM-pf complex for DBS therapy, based on the fact that electrical stimulation of the CM-pf complex is known to induce incremental recruiting and an augmenting response of the EEG with low-frequency stimulation, and EEG desynchronization with high-frequency stimulation [1, 4]. Major series of reports [6, 7] on recovery from the PVS have indicated that recovery of consciousness from a posttraumatic PVS is unlikely after 12 months, and recovery from a nontraumatic PVS after 3 months is exceedingly rare. Our 8 cases who emerged from the PVS included 6 cases of cerebrovascular injury and 2 cases of traumatic injury. We stress that chronic DBS therapy may be useful for allowing patients to emerge from a PVS, if the candidates are selected according to neurophysiological criteria. The above 8 cases were all in a bedridden state even after emerging from the PVS. However, we have experienced so-called minimally conscious state (MCS) [2] patients at 3 months after brain injury who recovered completely and became able to speak correctly and walk without help after DBS therapy. Clinical application of DBS therapy for the MCS is thus to be more expected when considering the physiologically disabled state in patients who are recovering from the PVS.

The Multi-Society Task Force on PVS [6, 7] summarized several papers and reported that 33% of adults in a vegetative state had died within 1 year after sustaining traumatic brain injury, and 53% of those after a non-traumatic injury had died within 1 year. Further, the mortality rate for adults in a PVS after acute brain injury is 82% at 3 years and 95% at 5 years, and survival beyond 10 years is unusual. The mortality rate of our DBS-treated patients was 43% at 3 years, 71% at 5 years, and 81% at 10 years. The fact that 4 of 21 cases (19%) survived for over 10 years should be considered as a specific effect of the DBS therapy. Among the 4 cases who survived for over 10 years, 3 cases emerged from the vegetative state and 1 case remained in a vegetative state. The finding that 3 of 8 cases (38%) who emerged from the PVS and 1 of 13 cases (8%) who remained in a PVS survived for over 10 years indicates that recovery from the vegetative state is important for long-term survival even in a bedridden state. The difference in survival period at over ca. 6 years was significant between the recovered and non-recovered cases.

Our results demonstrate that DBS therapy can be effective in allowing some patients to emerge from the vegetative state, and can lead to long-term survival. Considering the existence of a physiologically disturbed condition, a special rehabilitation program for prolonged coma may be necessary. The MCS is characterized by inconsistent but clearly discernible behavioral evidence of consciousness and can be distinguished from coma and a vegetative state by documenting the presence of specific behavioral features not found in either of these conditions. MCS is often transient but may also exist as a permanent outcome. We expect that DBS therapy used for the treatment of MCS patients will help to accelerate recovery from this state.

Acknowledgment

This work was supported by grants from the Ministry of Education, Science and Culture of Japan (B09470302 and A12307029).

References

1. Dempsey EWA, Morison RS (1942) A study of thalamocortical relation. Am J Physiol 135: 291–292
2. Giacino JT, Ashwal S, Childs N, Cranford R, Jennett B, Katz DI *et al* (2002) The minimally conscious state: Definition and diagnostic criteria. Neurology 58: 349–353
3. Jennett B, Plum F (1972) Persistent vegetative state after brain damage. Lancet 1: 734–737
4. Jasper HH, Naquet R, King LE (1955) Thalamocortical recruiting responses in sensory receiving areas in the cat. Electroencephalogr Clin Neurophysiol 7: 99–114
5. Katayama Y, Tsubokawa T, Yamamoto Y, Hirayama T, Miyazaki S, Koyama S (1991) Characterization and modification of brain activity with deep brain stimulation in a persistent vegetative state. Pacing Clin Electrophysiol 14: 116–121
6. The Multi-Society Task Force on PVS (1994) Medical aspects of the persistent vegetative state. (First of two parts). N Engl J Med 330: 1499–1508
7. The Multi-Society Task Force on PVS (1994) Medical aspects of the persistent vegetative state. (Second of two parts). N Engl J Med 330: 1572–1579
8. Tsubokawa T, Yamamoto Y, Katayama Y, Hirayama T, Maejima S, Moriya T (1990) Deep brain stimulation in a persistent vegetative state: Follow-up results and criteria for selection of candidates. Brain Injury 4: 315–327
9. Tsubokawa T, Yamamoto Y, Katayama Y (1990) Prediction of the outcome of prolonged coma caused by brain damage. Brain Injury 4: 329–337
10. Yamamoto T, Katayama Y, Oshima H, Fukaya C, Kawamata T, Tsubokawa T (2001) Deep brain stimulation therapy for a persistent vegetative state. Acta Neurochir (Wien) [Suppl] 79: 79–82

Correspondence: Takamitsu Yamamoto, M.D., Ph.D., Department of Neurological Surgery and Division of Applied System Neuroscience, Nihon University School of Medicine, 30-1 Ohyaguchi Kamimachi, Itabashi-ku, Tokyo 173-8610, Japan. E-mail: nusmyama@med.nihon-u.ac.jp

Acta Neurochir (2003) [Suppl] 87: 19–21

Clinical study on effect of HBO plus electric stimulation on treatment for the vegetative state

K. Xie[1] and **P. Wang**[2]

[1] Emergency Center, Affiliated Nanjing Drum Tower Hospital, Medical School of Nanjing University, Nanjing, P.R. China
[2] Nanjing University Medical School, Affiliated Gu-Lou Hospital, P.R. China

Summary

We proposed neurological diagnostic criteria and new scoring system for the vegetative state (VS). According to these criteria and scoring system, we examined the effects of hyperbaric oxygen (HBO) treatment in 130 cases of VS. As compared with non-HBO treatment, HBO treated patients showed statistically significant recovery from VS.

Keywords: Vegetative state; hyperbaric oxygen therapy; scoring system.

Introduction

The vegetative state is a clinical condition involving complete unawareness of self and environment, accompanied by sleep-wake cycles at more than one month after brain damage as initial insult. In this connection, two consecutive, national conferences were being held in June 1996 and November 2001.

The Glasgow Coma Scale (GCS) is a popular and helpful tool to assess the acute phase of patients with cerebral damage but lacks precision in chronic illness. A new and simple neurologically based grading system was developed during the November 2001 meeting in Nanjing. We proposed the following neurological diagnostic criteria for the vegetative state:

(1) No evidence of awareness of self or environment and inability to interact with others
(2) Spontaneous eye opening or eye opening in response to stimulation
(3) Intermittent wakefulness manifested by the presence of sleep-wake cycles
(4) Non-purposeful pursuit by eye movement
(5) No evidence of language comprehension or expression
(6) Spontaneous respiration and normal BP
(7) Sufficiently preserved hypothalamic and brainstem autonomic functions

The new scoring system has convincingly been applied to patients in the convalescent stage of VS in accordance with the November 2001 meeting. The scoring system for evaluating VS effectiveness has two dimensions, namely: (1) Roman numeral I, II, III: indicates whether or not a patient emerged from the vegetative state (primarily communication capacity) and (2) other neurological findings improving quality of life.

(1) Whether or not a patient emerged from the vegetative state (extent of recovery):
  I. persistent vegetative state: loss of cognitive functions and inability to follow directions (not including aphasia).
  II. initial emergence from VS: patient can follow directions or respond to simple questions.
  III. emergence from VS: patient can follow directions and responds to questions fairly well.
(2) Other neurological findings improving quality of life:

Movement	Score	Movement	Score
Limb's motion		*eye movement*	
– No voluntary movement	0	– None	0
– Meaningless motion	1	– Eye pursuit	1
– Command response	2	– Purposeful	2

Movement	Score	Movement	Score
Emotional response		*EEG*	
		– Isoelectric record	0
– None	0	– Delta or Theta rhythms	1
– Slight	1		
– Responses	2	– Alpha or Beta rhythms	2
Food intake		*SEP*	
– Tube feeding	0	– N_{20} disappearance	0
– Swallowing	1		
– Chewing + swallowing	2	– N_{20} prolonged latency	1
		– N_{20} normal	2

The neurophysiological changes in VS patients vary from 0 to 12 points. A 0 to 1 score means no effectiveness; score ranging from 2 to 3 means little effectiveness; 4 to 7 points: effectiveness, and more than 8 points indicates significant effectiveness.

The recording system is as follows: I; II; III plus 0~12 points.

For example, II6 indicates that the patient has initially emerged from VS and other physiological functions are improved. Hence, the communication capacity is the most important fact.

Materials and methods

Eligible patients included men and non-pregnant women, 19 to 76 years of age (mean age 47.5 years) who had suffered traumatic brain injury (101 cases), brain hemorrhage (8 cases), anesthesia accidents (4 cases), gas intoxication (5 cases), electric shock (3 cases), tried to commit suicide by hanging (2 cases), and others (2 cases). Among all patients, 93 cases had a haematoma removed, underwent decompression operation, or cardiac-pulmonary resuscitation. In all patients CT scan and MRI examination were performed and after first aid and comprehensive treatments had developed into PVS before being admitted. During the hospital stay, complications were as follows: 64 cases (49.23%) of pulmonary infections, 18 cases (13.8%) of hemorrhage of irritable ulcers, 13 cases (10%) of epilepsy, 11 (8.46%) cases of bedsore, 6 cases (4.6%) of urinary infections, 4 cases (3.1%) of keratitis. To qualify, all patients accorded with the criteria.

Treatment

1. *HBO*: All 130 patients received HBO treatment in large HBO cabins with compressed air and mask. Patients who underwent tracheotomy used a connector to inhale oxygen. We devised different treatment schemes according to the different etiology, course of disease, patients' condition and γ CBF, EEG, CT and MRI, cerebral evoked potential. Usually, the air is pressurised to 0.25 Mpa (2.5 ATA), and patients use mask to inhale oxygen for 30 minutes twice, and inhale air for 10 minutes intermittently (once or twice per day). 30 to 40 times constitutes a single therapeutic course. If the patient is in a severe state and has unstable vital signs, the air pressure is modulated to 0.16~0.20 Mpa, 2 to 3 times per day. This group of patients received no more than 210 times of HBO treatment and no less than 30 times with a mean of 120 times.
2. *Cervical Stimulation of Spinal Cord (CSCS)*: Among the 130 cases, 11 cases were given CSCS (10 men and 1 woman). Their ages ranged from 21 to 63 years, and duration of disease was 58 to 568 days. There were 6 cases of severe brain injury, 4 cases of anoxemic encephalopathy, and one case of cerebral vascular malformation.
3. *Secondary therapy*:
   3.1 Drug therapy
   To improve microcirculation and cerebral metabolism.
   3.2 Acupuncture and masotherapy
   It is desirable to apply acupuncture to regain consciousness and restore mental clarity for all patients in VS following treatment with masotherapy. According to traditional Chinese medical literature, acupuncture and masotherapy can improve blood circulation and relieve stasis.
   3.3 Affectionate nursing and family care is important.
   3.4 Rehabilitation of nerves and limbs as soon as possible.

Table 1. *Etiology of PVS*

Traumatic			Nontraumatic					
Traffic accidents	Beating	Others	Anesthesia accidents	Brain hemorrhage	Electric shock	Intoxication	Suicide	Others
101	4	1	4	8	3	5	2	2

Results

Table 2. *The curative effects of 130 cases of patients are as follows*

Curative effects	Score	Cases	%
II (excellent)	12~18	39	30.0
II (good)	6~11	56	43.1
I (improved)	2~5	14	10.7
I	without change	21	16.2

Table 3. *Frequency effect of HBO treatment on PVS*

Group	Frequency	Case	Curative effects (case)	
			Almost recovered	Efficiency
A	<60	28	9	32%
B	60~100	44	27	61.4%†*
C	>100	58	3	5.2%
Total		130	39	30%

† Shows B vs. A, $P < 0.05$; * shows B vs. C, $P < 0.01$.

Table 4. *Effect of HBO and non-HBO treatment on PVS in China and outside China*

Group	Unit	Cases	Efficiency (%)
A.	Nanjing Military Surgeon's College	98	76.4
HBO Treatment*	Lanzhou First Hospital of PLA	41	80.48
	Xiangya Hospital of Human Medical Univ.	60	86.7
	Anhui General Hospital of Armed Police	11	62.7
B.	Heilongjiang Rehabilitation Hospital	104	34.6
Non-HBO Treatment	ICDB (International Coma Database)	140	42.0
	TCDB (Traumatic Coma Database)	84	52.0
	Heilongjiang Rehabilitation Hospital	30	54.6

* Shows A vs. B, $P < 0.01$.

Correspondence: Xie Kangmin, M.D., Nanjing Gulou Hospital Consultant Center, Nanjing, 210008, P. R. China. E-mail: kangmin_xie@sina.com

Acta Neurochir (2003) [Suppl] 87: 23–26

Effects of musicokinetic therapy and spinal cord stimulation on patients in a persistent vegetative state

R. Noda[1], **Y. Maeda**[2], and **A. Yoshino**[3]

[1] Arts Planning Department, Osaka University of Arts, Osaka, Japan
[2] Department of Neurosurgery, Ishikiriseiki Hospital, Osaka, Japan
[3] Department of Neurological Surgery, Nihon University School of Medicine, Tokyo, Japan

Summary

We developed a method of musicokinetic therapy (MKT), employing a trampoline with live music performance (saxophone or electric piano), in an attempt to improve the clinical condition of patients in a persistent vegetative state (PVS). As an initial step for assessing the effect of MKT on PVS, we analyzed the changes in PVS score (range: 0–30) after MKT, which was continued for 3 months, in a consecutive series of 26 patients. These patients fulfilled the definition of PVS adopted by the Multi-Society Task Force on PVS. In this series, 7 patients were being treated by spinal cord stimulation at the same time. We, therefore, also had an opportunity to examine the effect of spinal cord stimulation on PVS. A greater or lesser improvement in PVS score (post-MKT score − pre-MKT score, mean ± SD: 8.27 ± 5.52) was observed in all patients except one. Among 12 patients who had been in PVS for 1 year or more before the initiation of MKT, 7 patients (56.3%) demonstrated improvement of their PVS score by 5 or more, and 4 patients (33.3%) reached a post-MKT score of greater than 20. The condition defined as PVS can never be scored better than 20. Since it is commonly felt that spontaneous improvement rarely occurs if PVS has continued for more than 6 months, the improvement after MKT appears to be better than that which could be observed spontaneously. The improvement in PVS score was often noted in patients with brain damage caused by trauma or SAH. There was no significant difference in improvement of the PVS score between patients who were treated by spinal cord stimulation and those who were not. Although the present study did not directly prove an effect of MKT on PVS, because no controls were involved, the results were consistent with the hypothesis that MKT is useful for improving the clinical condition of patients in PVS, especially those with severe brain damage caused by trauma or SAH.

Keywords: Musicokinetic therapy; persistent vegetative state; brain damage; spinal cord stimulation.

Introduction

As a result of progress in intensive medical treatment, many patients who would previously have died now survive. Many of the survivors, however, remain in a state of prolonged coma and 30–40 percent of them fall into a persistent vegetative state (PVS) [2, 10, 12]. The term PVS was coined by Jennett and Plum in 1972 to describe the condition of patients with severe brain damage in whom coma has evolved to a state of wakefulness without detectable awareness [3]. Recovery in awareness rarely occurs if PVS continues for more than 6 months [4, 11]. We developed a method of musicokinetic therapy (MKT) under the hypothesis that musical and kinetic stimulation, which are given synchronously and rhythmically, could help patients in PVS to regain awareness through activation of brain function [5, 6]. As an initial step for assessing the effect of MKT on PVS, we evaluated the changes in clinical condition after MKT, which was continued for 3 months, in a consecutive series of 26 patients in PVS. In this series, 7 patients were being treated by spinal cord stimulation at the same time. We, therefore, also had an opportunity to examine the effect of spinal cord stimulation on PVS.

Methods

The patients in the present study were consecutively treated by MKT at the Ishikiriseiki Hospital, Osaka, Japan, between October, 1997 and November, 2001. Their ages ranged from 14 to 72 (median, 38.5) years; 18 patients were male, and 8 were female. The causes of the brain damage were head trauma (12 cases), aneurysmal subarachnoid hemorrhage (SAH: 9 cases), other cerebrovascular accidents (3 cases), and anoxic encephalopathy (2 cases). As mentioned, 7 of them were being treated by spinal cord stimulation with a chronically implanted stimulation system.

The method of MKT was as described previously [5, 6]. In brief, a pair of caregivers supports the patient in a sitting position and give stimulation through vertical motions on a trampoline. At the same time, live music using a saxophone or electric piano is played in synchronization with the vertical motion. Following vertical motions for 5 to 10 minutes, the patient is allowed to lay on the trampoline and listen to relatively slow and soft music for 3 to 5 minutes while receiving massage of the extremities. These procedures are repeated and each session is continued for 30 minutes. Such MKT was initiated at varying periods after brain damage ranging from one month to 4 years and 9 months (median, 327 days), and performed once a week for 3 months or more (3–12 months). The effect of the MKT was evaluated on the basis of the PVS score (Table 1) as defined by the Society for Treatment of Coma, Japan [9]. The PVS score ranges from 0 to 30. The condition defined as PVS can never be scored better than 20. The PVS score was taken before (pre-MKT score) and 3 months after (post-MKT score) the initiation of MKT.

Results

The PVS score was significantly improved after MKT (pre-MKT score vs. post-MKT score: 10.50 ± 4.74 vs. 18.77 ± 7.68; Wilcoxon's signed rank test, $p < 0.0001$). A greater or lesser improvement in PVS score (post-MKT score – pre-MKT score) was observed in all patients except one who demonstrated no change in PVS score after MKT. There were 12 patients who had been in PVS for 1 year or more before the initiation of MKT (Table 2). Among them, 7 patients (56.3%) demonstrated improvement of their PVS score by 5 or more, and 4 patients (33.3%) reached a post-MKT score of greater than 20. These 4 patients were ones who had brain damage caused by trauma or SAH. There was no patient who demonstrated a decrease in PVS score or died during MKT. The post-MKT score or improvement of the PVS score was not related to either age or sex. There was no difference in improvement of the PVS score between the patients who were treated by spinal cord stimulation and those who were not.

The improvement in the PVS score appeared to be dependent on the cause of brain damage (Table 3). Ten (83.3%) of the 12 patients with brain damage caused by trauma and 7 (77.8%) of the 9 patients with brain damage caused by SAH demonstrated improvement of their PVS score by 5 or more after MKT. Eight (66.7%) and 5 (55.6%) of the patients with brain damage caused by trauma or SAH reached a post-MKT score of greater than 20, respectively. In contrast, only one (33%) of the 3 patients with brain damage caused by cerebrovascular accident other than SAH and none of the 2 patients with anoxic brain damage demonstrated improvement of their PVS score by 5 or more after MKT. None of the patients with brain damage caused by cerebrovascular accident other than SAH or with anoxic brain damage reached a post-MKT score of greater than 20.

Table 1. *PVS scale (Society for Treatment of Coma, Japan. 1997)*

State scale (full score, 10 points)	
✓ Sufficient spontaneous respiration	
✓ Spontaneous eye opening	
✓ Sleep-wake cycles, even if only slight	
✓ Spontaneous movements of the oral cavity and pharynx	
✓ Spontaneous movements of the extremities	
✓ Spontaneous change in the facial expression muscles	
✓ Spontaneous vocalization	
✓ Interest in environment	
✓ Voluntary purposeful body movement	
✓ Spontaneous consistent ability to verbalize coherently (speech)	
Reaction scale (full score, 20 points)	
Eye opening responses	point
Eye opening and closing in response to verbal stimuli	4
Eye opening in response to verbal stimuli	3
Eye opening in response to strong stimuli such as loud noises or being shaken bodily	2
Eye opening in response to painful stimuli	1
None	0
Limb motoric responses	
Sustained, reproducible and purposeful limb movement in response to verbal stimuli	4
Sustained, reproducible and purposeful limb movement in response to painful stimuli	3
Limb withdrawal in response to painful stimuli	2
Limb flexion in response to painful stimuli	1
None	0
Emotional responses	
Abundant emotional expression in response to various stimuli	4
Reaction of the facial expression muscle to various non-painful stimuli	3
Reaction of the facial expression muscle to painful stimuli	2
Reaction of the autonomic nervous system to painful stimuli	1
None	0
Head and eye responses	
Sustained, reproducible and purposeful eyeball movements and fixation in response to visual stimuli	4
Eye closure (blink response) to visual threatening stimuli	3
Brief eye tracking in response to visual stimuli	2
Brief orienting response of the head and/or eyeball to visual and/or auditory stimuli	1
None	0
Vocalization responses	
Responding with appropriate conversations to various stimuli	4
Responding with simple sentences to various stimuli	3
Responding with words to various stimuli	2
Vocalization in response to various stimuli	1
None	0

The PVS score is adopted the total points of State Scale and Reaction Scale.

Table 2. *Timing of the initiation of MKT and changes in the PVS score after MKT*

Timing of MKT after brain damage	n	Improvement in the PVS score > 5	Post-MKT PVS score > 20
<6 months	9	9 (100.0%)	9 (100.0%)
6–12 months	5	2 (40.0%)	0
>12 months	12	7 (56.3%)	4 (33.3%)

Table 3. *Cause of brain damage and changes in the PVS score after MKT*

Cause of brain damage	n	Improvement in the PVS score > 5	Post-MKT PVS score > 20
Trauma	12	10 (83.3%)	8 (66.7%)
SAH	9	7 (77.8%)	5 (55.6%)
Cerebrovascular accident other than SAH	3	1 (33.3%)	0
Anoxia	2	0	0

Discussion

MKT is designed to activate the central nervous system efficiently by using environmental stimuli which are meaningful for patients. The motion of the patient produced by the trampoline stimulates various sensorimotor pathways efficiently. When the motion is synchronized with music, it may become meaningful if the music is felt familiar to the patient [5, 6]. It is hypothesized that such stimuli from the environment can help in the restoration of awareness. The results of the present study are consistent with the hypothesis that MKT is useful for improving the clinical condition of patients in PVS. However, definite conclusions cannot be drawn regarding the precise effect of MKT on PVS, because the present study did not involve controls and the extent of spontaneous improvement could not therefore be evaluated. It is commonly thought that spontaneous improvement rarely occurs if PVS has continued for more than 6 months [4, 11]. In the present study, however, noticeable improvement of clinical condition (improvement in PVS score > 5) was induced in more than half of patients who had been in PVS for more than 1 year. Furthermore, one third of them reached a clinical condition which was clearly better than the condition defined as PVS (PVS score > 20). This implies that MKT may in fact have induced the improvement of clinical condition.

The present study failed to demonstrate an effect of spinal cord stimulation on the PVS score. This does not mean that spinal cord stimulation is completely useless for improving the clinical condition of patients in PVS: the present number of patients treated by spinal cord stimulation was relatively small, and patients who had demonstrated good improvement by spinal cord stimulation had not been enrolled on the MKT program.

The present data suggest that patients in PVS caused by trauma or SAH possess a greater potential for improvement in their clinical condition through MKT, as compared to patients in PVS caused by cerebrovascular accident other than SAH or anoxia. This is in agreement with previous studies which have shown that outcome of patients in PVS may depend on the cause of brain damage [1, 7, 8]. The Multi-Society Task Force on PVS [7, 8] indicated that outcome is better in patients with brain damage caused by trauma, as compared to patients with non-traumatic brain damage, especially those with anoxic encephalopathy. With appropriate controls and randomization, future studies on patients with brain damage caused by trauma or SAH should be able to reveal more clearly the exact effect of MKT on PVS.

References

1. Council on Scientific Affairs and Council on Ethical and Judicial Affairs (1990) Persistent vegetative state and decision to withdraw or withhold life support. JAMA 263: 426–430
2. Groswasser Z, Sazbon L (1990) Outcome in 134 patients with prolonged post-traumatic unawareness: I. Parameters determining late recovery of consciousness. J Neurosurg 72: 75–80
3. Jennett B, Plum F (1972) Persistent vegetative state after brain damage. Lancet 1: 734–737
4. Katayama Y, Tsubokawa T, Yamamoto T, Hirayama T, Miyazaki S, Koyama S (1991) Characterization and modification of brain activity with deep brain stimulation in a persistent vegetative state. Pacing Clin Electrophysiol 14: 116–121
5. Maeda Y, Koba K, Kajiki M, Noda R (2001) Effects of musical exercise therapy on 26 patients with prolonged disturbance of consciousness. Proceedings of the 10th Annual Meeting of the Society for Treatment of Coma, vol 10: pp 113–121
6. Noda R (2000) Stimulation of the five senses for patients with prolonged consciousness disturbance: effects of musical exercise therapy on prolonged consciousness disturbance. Proceedings of the 9th Annual Meeting of the Society for Treatment of Coma, vol 9: pp 9–21
7. The Multi-Society Task Force on PVS (1994) Medical aspects of the persistent vegetative state (First of two parts). N Engl J Med 330: 1499–1508
8. The Multi-Society Task Force on PVS (1994) Medical aspects of the persistent vegetative state (Second of two parts). N Engl J Med 330: 1572–1579
9. The Society for Treatment of Coma, Japan (1997) Scoring system for PVS. Nikkenkagaku Co Ltd, Tokyo

10. Tsubokawa T, Yamamoto T (1996) Deep brain stimulation in the persistent vegetative state. In: Gildenberger PL, Tasker RR (eds) Textbook of stereotactic and functional neurosurgery. McGraw-Hill New York, pp 1979–1986
11. Tsubokawa T, Yamamoto T, Katayama Y (1990) Prediction of the outcome of prolonged coma caused by brain damage. Brain Injury 4: 329–337
12. Yamamoto T, Katayama Y, Oshima H, Fukaya C, Kawamata T, Tsubokawa T (2001) Deep brain stimulation therapy for a persistent vegetative state. Acta Neurochir (Wien) [Suppl] 79: 79–82

Correspondence: Atsuo Yoshino, M.D., Ph.D., Department of Neurological Surgery, Nihon University School of Medicine, 30-1 Oyaguchi-Kamimachi, Itabashi-ku, Tokyo 173-8610, Japan. E-mail: ayoshin@med.nihon-u.ac.jp

Clinical application of the drug pump for spasticity, pain and restorative neurosurgery

Acta Neurochir (2003) [Suppl] 87: 29–35

Intrathecal drug delivery 2002

M. S. Turner

Pediatric Neurosurgery, Indianapolis Neurosurgical Group, Inc., IN, USA

Summary

Background. Intrathecal drug delivery has been used clinically since the 1970's. Significant clinical advances have been made combining new technology with pharmacology and surgery. Continuous infusion of medication for both analgesia and spasticity has become a part of the armamentarium for specialists in these areas. Significant recent advances in technology promise further enhancements and improvements for intrathecal therapy.

Methods. A review of the literature combined with 20 years personal experience with intrathecal drug delivery.

Findings/discussion. Intrathecal therapy has established a role in the treatment of malignant pain, benign pain and severe spasticity. Significant literature and the current state of practice in the United States are reviewed. Recent therapeutic enhancements are discussed, and a wish list of future technological enhancements presented.

Keywords: Intrathecal; baclofen; morphine; infusion pump.

Introduction

Intrathecal delivery of medications was initially introduced in 1975. The attraction of intrathecal drug delivery is its ability to deliver the drug to the site of action and bypass the blood brain barrier. A drug molecule must reach its site of action to be effective. Intrathecal drug delivery allows a drug molecule to be delivered to or very near its site of action in the spinal cord. The delivery of the drug directly to its site of action has several very attractive benefits:

- Greater therapeutic effect is possible because of the ability to deliver larger amounts of drug molecules to the site of action without systemic side effects due to the drug's action at other sites.
- The lack of systemic distribution eliminates side effects due to the drug acting at another site in the body.
- Intrathecal delivery bypasses the blood brain barrier.
- The amount of drug needed to get a clinical effect is significantly reduced, as the drug is not distributed throughout the body.

Chronic continuous infusion of intrathecal medication with implanted catheters and infusion pumps began in the 1980's. Early intrathecal administration included antibiotics, antifungals, steroids and cancer chemotherapy drugs. Administration was delivered as a bolus through a lumbar puncture. Initial treatment was primarily for fungal and bacterial meningitis, carcinomatous meningitis, and multiple sclerosis. In recent years continuous intrathecal drug delivery has proven effective in pain and movement disorders.

History

The identification of opiate receptors in nervous tissue in 1970 revealed the potential for opiate intrathecal drug delivery [58]. Several years later the opiate receptor was identified in the spinal cord [7]. By 1979 intrathecal administration of morphine was reported to give substantial relief of pain in patients with cancer pain [70]. Intrathecal and epidural use of morphine progressed rapidly in the 80's and 90's. The Food and Drug Administration (FDA) cleared use of morphine administered by a continuous non-programmable infusion pump, (Infusaid®, Arrow International;) in 1982. The programmable Medtronic Synchromed system was released in 1988, and cleared by the FDA for its use for intrathecal administration of morphine for cancer pain and benign pain in 1991. Intrathecal treatment of spasticity began with the bolus injections of alcohol and phenol by lumbar puncture in 1970 [20]. Neurotoxicity limited its use to patients with significant deficits. The synthesis of baclofen, a gamma amino butyric acid (GABA) analogue allowed for the increased clinical administration of the intrathecal route for spasticity. In the early 1980's GABA receptors were identified in the spinal cord [59] and the effectiveness of baclofen on inhibition of spinal cord reflexes was demonstrated [8]. In 1984 Penn reported the use of intrathecal baclofen to relieve spasticity in spinal cord injury in humans [55]. The FDA cleared the continuous infusion of intrathecal baclofen for severe spasticity of spinal origin in June of 1992 and for severe spasticity of cerebral origin in June of 1996.

Current practice in the United States

Pain

Continuous infusion of intrathecal medication for pain relief has become common practice in the United States. It is frequently used on a temporary (2–3 days) basis, administered as an epidural infusion, with the opiate drugs morphine sulfate or fentanyl and administered for control of postoperative pain. The local anesthetic drug bupivicaine is also used for acute postoperative pain.

Continuous infusion of opiates on a long-term basis using implanted continuous infusion pumps currently includes the routine use of the opiates morphine, dihydromorphone, and fentanyl [29, 60]. Bupivicaine is used in continuous long-term infusion systems, especially in patients with neuropathic pain. The alpha-adrenergic drug clonidine is used effectively in patients with neuropathic pain, but far less commonly than bupivicaine. Baclofen has been reported to be useful in central deafferentation pain from spinal cord injury and stroke [26, 48, 65, 72], and in complex regional pain syndromes [67, 73].

Intrathecal administration of opiates was initially approved for pain in cancer patients [15, 37]. It is now widely used in benign pain; primarily failed back syndrome and complex regional pain syndrome [12, 13]. Limited or investigational intrathecal use of other medications for pain includes the trial of SNX 111 for cancer pain [34, 56], octreotide [51] adenosine [11] and use of ketamine in selected patients [29, 64].

Many patients with chronic pain have combinations of medication infused, including morphine with bupivicaine, morphine with clonidine, other opiates with either clonidine or bupivicaine. Other drug combinations are also used [22, 29, 50].

Potential complications

Opiate tolerance can develop with intrathecal administration and can present a problem in patients with benign pain. A complication associated with chronic intraspinal administration of opiates is the development of catheter tip granulomas. They appear to be only with the higher concentration of medication, and remain rare, although the exact incidence is unknown [17, 40].

Intensive care

Intrathecal baclofen has been successfully used for treatment of rigidity from tetanus [14, 21, 62]. This is usually accomplished with temporary catheter systems. Similarly short term infusion of baclofen has been used early in traumatic brain injury [10, 19] to abolish the severe posturing and dysautonomia. Early studies have suggested potential for intrathecal nitroprusside for cerebral vasospasm [66].

Spasticity and movement disorders

The FDA cleared the intrathecal administration of baclofen with the Synchromed® System (Medtronic) in 1992 for severe spasticity of spinal origin. This includes its use in patients with both spinal cord injury [9] and multiple sclerosis [18, 46, 53, 54, 57]. In the United States spasticity treatment centers offer combinations of therapies individualized for each patient [27, 33]. The therapies include intrathecal baclofen (ITB), often with combinations of physical therapy, casting, motor point blocks, botulinum toxin injection, and in selected children with cerebral palsy, and selective dorsal rhizotomy (SDR). The FDA cleared chronic intrathecal use of baclofen in patients with severe spasticity of cerebral origin in 1996. This category includes patients with traumatic brain injury [3, 44], cerebral palsy [1, 3], and stroke [43]. Intrathecal baclofen (ITB), has shown to effectively improve the quality of life in patients with spasticity [47].

ITB has been used effectively in many patients with cerebral palsy [2, 6, 16, 27]. Many of the more severely involved patients with spasticity from cerebral palsy also have dystonia. It became apparent in these children that the dystonic symptoms improved significantly with ITB. This is in contrast to the experience in children with spasticity and dystonia that underwent SDR, in which the dystonic symptoms failed to respond. Some early reports suggested that dystonia did not respond to ITB [24]. However numerous centers reported positive outcomes with the use of ITB in patients with cerebral palsy and dystonia [3, 5, 24, 30, 45, 52]. Our center has had success with the use of ITB in patients with dystonia from cerebral palsy, post traumatic dystonia and generalized dystonia. We have experienced success in selected patients with cerebral palsy and athetosis. Current clinical studies are underway in several centers to better identify those patients with movement disorders that would benefit from ITB.

Future

The FDA cleared the use of ITB for traumatic brain injury (TBI) one year after injury. Our center has been using ITB (much earlier), as early as one month post injury. These patients present with severe spasticity and many have dysautonomia. ITB has been an effective treatment for those patients with post traumatic or post anoxic dysautonomia [10, 19], eliminating the high fevers, tachycardia, tachypnea, and opisthotonic posturing. ITB rapidly eliminates the severe decorticate posturing seen in severe brain injury and allows the patient to be weaned from sedating oral medications and to either progress to active rehab, or to be suitable for care at home. The elimination of the spasticity allows the patient to be out of bed in a chair, or mobilized out of the hospital in a wheelchair as opposed to a cart. This significantly decreases the work required to bathe and do daily hygiene for these patients. The mixing early with late use of ITB in dysautonomia, even several years post injury has been effective in stopping symptoms. All of our patients with early use of ITB (with an implanted pump) still benefit form ITB in reduction of tone, and several of our patients are more that 5 years post implant.

Early use of ITB in brain injury appears to decrease the time of hospitalization, and the cost of care for these individuals, often potentially saving thousands of dollars [61]. Unfortunately, these patients represent a very diverse group, and both the length of stay and cost for treatment are influenced by many factors. It will require a RCT (Randomized, {prospective, masked} Clinical Trial) to prove the cost effectiveness of early ITB.

Pharmacokinetics

Initial studies by Penn and Kroin [38] demonstrated a 4-1 ratio of drug concentration between the lumbar and cervical regions. Early experience with quadriplegic spinal cord injured patients showed that the reduction in spasticity in the lower extremities was greater than the upper extremities. The effective treatment of dystonia and athetosis with ITB has depended on the placement of catheters higher in the spinal canal. Neurosurgeons then began to move the tip of the catheter cephalad in the spinal column [28, 32]. They hoped to get an increased effect on the upper extremities by increasing the concentration of drug in the cervical region (Fig. 1). This has proven clinically effective. The improved upper extremity effect could be achieved without increasing sedation or other side effects.

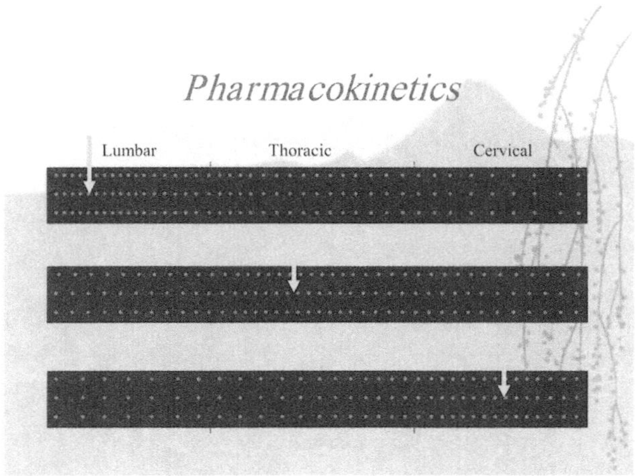

Fig. 1. Pharmacokinetics

Patients with movement disorders are a diverse group with many differing presentations, etiologies, syndromes and clinical problems. The application of intrathecal baclofen is still in development, and the ideal patients have not yet been identified.

More experienced clinicians are now using ITB to assist ambulatory and less severely involved patients, especially in the treatment of stroke.

The current pump (Synchromed EL®, Medtronic) is approved for subcutaneous implantation on the abdominal wall. In slender individuals this can be cosmetically unattractive, and further degrade an already poor body image. More involved cerebral palsy patients may get around by "commando" crawling on their abdomen, or rolling, and a large subcutaneous pump impairs mobility. Other patients with poor wound healing have been prone to wound breakdown; leading to infection and removal of the pump. Some clinicians have begun implanting the pump in the sub-

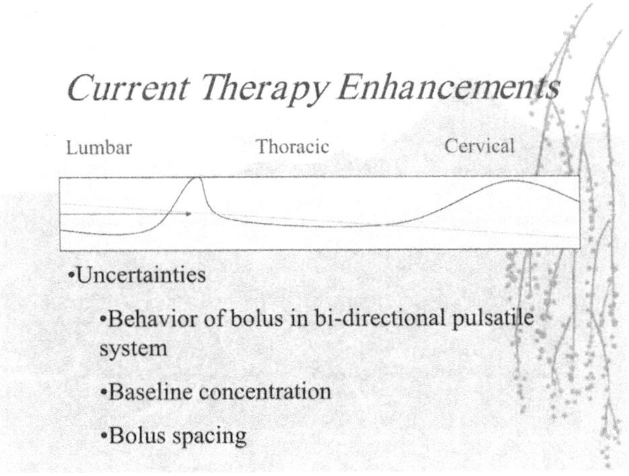

Fig. 2. Current therapy enhancement

fascial space, placing it beneath the external oblique muscle and the rectus abdominus fascia. This has served to both to decrease the infection rate, and provide better patient satisfaction [36].

A group of patients receiving ITB have reported that although they are experiencing a reduction in their spasticity, the improvement is not as marked as during their initial test dose. The initial test dose was administered as a bolus of medication. Clinicians have begun to vary the mode of drug administration in selected patients that fail to get the desired response, in spite of significant dose escalation. Instead of a continuous infusion, they administer a series of boluses, often 15–25% of their daily dose over 10 minutes every 4–6 hours, and report improvement in tone and decreased total daily dose (Fig. 2 Current therapy enhancement).

The future

The major cause of complications with the continuous infusion system has been directly associated with the catheters [23, 41, 69]. Unexpected pump failure accounts for less than 0.5% of reported complications (Personal communication, Medtronic, Inc.). Catheter problems include catheter extrusion, catheter disconnection, misplacement, kinking and cuts in the catheter. These problems are unfortunately technical and due to surgical technique. Medtronic analyzed adverse events from catheters as part of a 16 center Phase III study of intrathecal administration of a new compound. They found that the reported adverse events from catheter failure ranged from 6%–60% (center dependent). The development of catheter systems, with better connectors and anchor systems will help, but identification of best technique for catheter implantation and its adoption by implanters is essential to decreasing complications.

Many of the patients that are candidates for intrathecal drug delivery are significantly impaired and have great difficulties with transportation. Care of the patient with a continuous infusion pump is often limited by the necessity of the patient getting to a center for reservoir refills and reprogramming. The current pump is quite large and holds only 18 ml. of drug. The challenge for future pumps is to decrease size but hold more medication. Smaller pumps would allow implantation in other positions in the body, be less visible, and be implantable in smaller patients. Patient activated boluses could offer significant advantages to patients with pain, allowing the increase in medication for increases in pain, without the use of addition sedating oral or intravenous medication. The current pump also has very limited memory. Ideally the pump would keep a record of all changes to therapy, any patient activated therapy, catheter data including revisions and catheter length and contact phone numbers for the implanting and monitoring physicians. Preprogrammed safety parameters could alarm when programming parameters entered fall outside of established ranges. The pump could also do the complex calculations required when changing concentrations and do the appropriate programming. The ability for the pump to monitor certain physiologic parameters or patient entered data such as episodes of severe pain or spasms would allow more rational decisions on dosage adjustment.

The use of several medications simultaneously is common in pain patients. Dosage adjustments are complex and limited by the different therapeutic ranges of the combinations of medication. A dual or triple pump or dual/triple reservoir pump would make care of those patients safer, easier, and may allow therapies using medications that are not compatible.

Several manufactures make non-programmable pumps that are less expensive and useful in patients with stable dose requirements. A less expensive pump with limited programming ability and memory could be very useful and cost effective for patients with malignant pain.

The pump programmer of the future should be able to perform all the calculations listed above, download the recorded data plus contain a database of all the

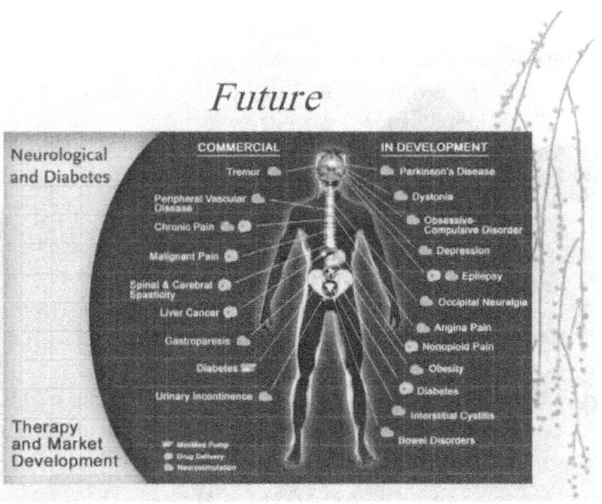

Fig. 3. Future

specific patient information on patients in therapy. This data should be transferable to a central computer in the physician's or clinic office to allow physicians and nurses access to vital data quickly when dealing with patient questions or potential problems. The potential to do remote telemedicine or Internet enhanced care of these patients also holds great appeal. The programmer of the future should also be lighter and smaller. A less complex patient programmer could also enhance therapy in selected patients and would be essential for telemedicine or e-medicine.

Future intrathecal drug delivery also depends on the development of new medications. Current research includes attempts to identify a non opiate analgesic, an analgesic for neuropathic pain, neurotropic growth factors or other substances for use in degenerative diseases such as amyotrophic lateral sclerosis (ALS) [49], multiple sclerosis (MS), brain injury, spinal cord injury [31], stroke, and cancer [35] (Fig. 3).

Delivery of medications to other areas of the brain, including along the walls of the ventricles, or even intraparenchymal could be useful. The current version of the Synchromed EL® pump (Medtronic) can deliver 50 micro liters per day and allow infusion of neurotransmitters, chemotherapy, antidepressants, or anticonvulsants through microcatheters directly to their site of action.

Great technological strides have been made in the 20th century to deliver drugs to their site of action in the brain and spinal column. Significant advances in computer utilization, manufacturing technology, and bio-technology represent a potential for future implantable drug delivery to be safer, clinician compliant, patient compliant, cost effective, and widely applicable in new patients and different diseases in the 21st century.

References

1. Albright AL (1996) Intrathecal baclofen in cerebral palsy movement disorders. J Child Neurol [Suppl] 1: S29–35
2. Albright AL (1996) Baclofen in the treatment of cerebral palsy. J Child Neurol 11(2): 77–83
3. Albright AL, Barron WB, Fasick MP, Polinko P, Janosky J (1993) Continuous intrathecal baclofen infusion for spasticity of cerebral origin. JAMA 270(20): 2475–2477
4. Albright AL, Barry MJ, Fasick P, Barron W, Shultz B (1996) Continuous intrathecal baclofen infusion for symptomatic generalized dystonia. Neurosurgery 38(5): 934–938
5. Albright AL, Barry MJ, Shafton DH, Ferson SS (2001) Intrathecal baclofen for generalized dystonia. Dev Med Child Neurol 43(10): 652–657
6. Armstrong RW, Steinbok P, Cochrane DD, Kube SD, Fife SE, Farrell K (1997) Intrathecally administered baclofen for treatment of children with spasticity of cerebral origin. Intensive Care Med 23(8): 896–9023
7. Atweh SF, Kuhar MJ (1977) Autoradiographic localization of opiate receptors in rat brain. I. Spinal cord and lower medulla. Brain Res 124: 53
8. Ault B, Evans R (1981) The depressant effect of baclofen on the isolated spinal cord of the neonatal rat. Eur J Pharmacol 71: 357–364
9. Azouvi P, Mane M, Thiebaut JB, Denys P, Remy-Neris O, Bussel B (1996) Intrathecal baclofen administration for control of severe spinal spasticity: functional improvement and long-term follow-up. Arch Phys Med Rehabil 77(1): 35–39
10. Becker R, Benes L, Sure U, Hellwig D, Bertalanffy H (2000) Intrathecal baclofen alleviates autonomic dysfunction in severe brain injury. J Clin Neurosci 7(4): 316–319
11. Belfrage M, Segerdahl M, Arner S, Sollevi A (1999) The safety and efficacy of intrathecal adenosine in patients with chronic neuropathic pain. Anesth Analg 89(1): 136–142
12. Bennett G, Deer T, Du Pen S, Rauck R, Yaksh T, Hassenbusch SJ (2000) Future directions in the management of pain by intraspinal drug delivery. J Pain Symptom Manage 20(2): S44–50
13. Bennett G, Serafini M, Burchiel K, Buchser E, Classen A, Deer T, Du Pen S, Ferrante FM, Hassenbusch SJ, Lou L, Maeyaert J, Penn R, Portenoy RK, Rauck R, Willis KD, Yaksh T (2000) Evidence-based review of the literature on intrathecal delivery of pain medication. J Pain Symptom Manage 20(2): S12–36
14. Boots RJ, Lipman J, O'Callaghan J, Scott P, Fraser J (2000) The treatment of tetanus with intrathecal baclofen. Anaesth Intensive Care 28(4): 438–442
15. Buchheit T, Rauck R (1999) Subarachnoid techniques for cancer pain therapy: when, why, and how? Curr Rev Pain 3(3): 198–205
16. Butler C, Campbell S (2000) Evidence of the effects of intrathecal baclofen for spastic and dystonic cerebral palsy. AACPDM Treatment Outcomes Committee Review Panel. Dev Med Child Neurol 42(9): 634–645
17. Coffey RJ, Burchiel K (2002) Inflammatory mass lesions associated with intrathecal drug infusion catheters: Report and observations on 41 patients. Neurosurgery 50(1): 78–86
18. Coffey JR, Cahill D, Steers W, Park TS, Ordia J, Meythaler J, Herman R, Shetter AG, Levy R, Gill B *et al* (1993) Intrathecal baclofen for intractable spasticity of spinal origin: results of a long-term multicenter study. J Neurosurg 78(2): 226–232

19. Cuny E, Richer E, Castel JP (2001) Dysautonomia syndrome in the acute recovery phase after traumatic brain injury: Relief with intrathecal Baclofen therapy. Brain Inj 15(10): 917–925
20. Davis R (1975) Spasticity following spinal cord injury. Clin Orthop 112: 66–75
21. Dressnandt J, Konstanzer A, Weinzierl FX, Pfab R, Klingelhofer J (1997) Intrathecal baclofen in tetanus: four cases and a review of reported cases. J Neurosurg 87(3): 409–410
22. Fanciullo GJ, Rose RJ, Lunt PG, Whalen PK, Ross E (1999) The state of implantable pain therapies in the United States: a nationwide survey of academic teaching programs. Anesth Analg 88(6): 1311–1316
23. Follett KA, Naumann CP (2000) A prospective study of catheter-related complications of intrathecal drug delivery systems. J Pain Symptom Manage 19(3): 209–215
24. Ford B, Greene P, Louis ED, Petzinger G, Bressman SB, Goodman R, Brin MF, Sadiq S, Fahn S (1996) Use of intrathecal baclofen in the treatment of patients with dystonia. Arch Neurol 53(12): 1241–1246
25. Francois B, Vacher P, Roustan J, Salle JY, Vidal J, Moreau JJ, Vignon P (2001) Intrathecal baclofen after traumatic brain injury: early treatment using a new technique to prevent spasticity. J Trauma 50(1): 158–161
26. Gatscher S, Becker R, Uhle E, Bertalanffy H (2002) Combined intrathecal baclofen and morphine infusion for the treatment of spasticity related pain and central deafferentiation pain. Acta Neurochir (Wien) [Suppl] 79: 75–76
27. Gormley Jr ME, Krach LE, Piccini L (2001) Spasticity management in the child with spastic quadriplegia. Eur J Neurol [Suppl] 5: 127–135
28. Grabb PA, Guin-Renfroe S, Meythaler JM (1999) Midthoracic catheter tip placement for intrathecal baclofen administration in children with quadriparetic spasticity. Neurosurgery 45(4): 833–836
29. Grabow TS, Derdzinski D, Staats PS (2001) Spinal drug delivery. Curr Pain Headache Rep 5(6): 510–516
30. Greene PE, Fahn S (1992) Baclofen in the treatment of idiopathic dystonia in children. Mov Disord 7(1): 48–52
31. Halter JA, Blight AR, Donovan WH, Calvillo O (2000) Intrathecal administration of 4-aminopyridine in chronic spinal injured patients. Spinal Cord 38(12): 728–732
32 Hugenholtz H, Nelson RF, Dehoux E (1993) Intrathecal baclofen – the importance of catheter position. Can J Neurol Sci 20(2): 165–167
33. Ivanhoe CB, Tilton AH, Francisco GE (2001) Intrathecal baclofen therapy for spastic hypertonia. Phys Med Rehabil Clin N Am 12(4): 923–938, viii–ix
34. Jain KK (2000) An evaluation of intrathecal ziconotide for the treatment of chronic pain. Expert Opin Investig Drugs 9(10): 2403–2410
35. Kerr JZ, Berg S, Blaney SM (2001) Intrathecal chemotherapy. Crit Rev Oncol Hematol 37(3): 227–236
36. Kopell BH, Sala D, Doyle WK, Feldman DS, Wisoff JH, Weiner HL (2001) Subfascial implantation of intrathecal baclofen pumps in children: technical note. Neurosurgery 49(3): 753–756; discussion 756–757
37. Krames ES (1999) Practical issues when using neuraxial infusion. Oncology (Huntingt) 13(5) [Suppl] 2: 37–44
38. Kroin JS, Ali A, York M, Penn RD (1993) The distribution of medication along the spinal canal after chronic intrathecal administration. Neurosurgery 33(2): 226–230
39. Kumar K, Kelly M, Pirlot T (2001) Continuous intrathecal morphine treatment for chronic pain of nonmalignant etiology: long-term benefits and efficacy. Surg Neurol 55(2): 79–86
40. Langsam A (1999) A case of spinal cord compression syndrome by a fibrotic mass presenting in a patient with an intrathecal pain management pump system. Pain 83(1): 97–99
41. Levin AB, Sperling KB, Penn RD, Gianino JM, York MM (1995) Complications associated with infusion pumps implanted for spasticity. Stereotact Funct Neurosurg 65(1–4): 147–151
42. Meythaler JM, DeVivo MJ, Hadley M (1996) Prospective study on the use of bolus intrathecal baclofen for spastichypertonia due to acquired brain injury. Arch Phys Med Rehabil 77(5): 461–466
43. Meythaler JM, Guin-Renfroe S, Brunner RC, Hadley MN (2001) Intrathecal baclofen for spastic hypertonia from stroke. Stroke 32(9): 2099–2109
44. Meythaler JM, Guin-Renfroe S, Grabb P, Hadley MN (1999) Long-term continuously infused intrathecal baclofen for spastic-dystonic hypertonia in traumatic brain injury: 1-year experience. Arch Phys Med Rehabil 80(1): 13–19
45. Meythaler JM, Guin-Renfroe S, Hadley MN (1999) Continuously infused intrathecal baclofen for spastic/dystonic hemiplegia: a preliminary report. Am J Phys Med Rehabil 78(3): 247–254
46. Meythaler JM, Steers WD, Tuel SM, Cross LL, Haworth CS (1992) Continuous intrathecal baclofen in spinal cord spasticity. A prospective study. Am J Phys Med Rehabil 71(6): 321–327
47. Middel B, Kuipers-Upmeijer H, Bouma J, Staal M, Oenema D, Postma T, Terpstra S, Stewart R (1997) Effect of intrathecal baclofen delivered by an implanted programmable pump on health related quality of life in patients with severe spasticity. J Neurol Neurosurg Psychiatry 63(2): 204–294
48. Middleton JW, Siddall PJ, Walker S, Molloy AR, Rutkowski SB (1996) Intrathecal clonidine and baclofen in the management of spasticity and neuropathic pain following spinal cord injury: a case study. Arch Neurol 53(12): 1241–1246
49. Ochs G, Penn RD, York M, Giess R, Beck M, Tonn J, Haigh J, Malta E, Traub M, Sendtner M, Toyka KV (2000) A phase I/II trial of recombinant methionyl human brain derived neurotrophic factor administered by intrathecal infusion to patients with amyotrophic lateral sclerosis. Amyotroph Lateral Scler Other Motor Neuron Disord 1(3): 201–206
50. Osenbach RK, Harvey S (2001) Neuraxial infusion in patients with chronic intractable cancer and noncancer pain. Curr Pain Headache Rep 5(3): 241–249
51. Paice JA, Penn RD, Kroin JS (1996) Intrathecal octreotide for relief of intractable nonmalignant pain: 5-year experience with two cases. Neurosurgery 38(1): 203–207
52. Paret G, Tirosh R, Ben Zeev B, Vardi A, Brandt N, Barzilay Z (1996) Intrathecal baclofen for severe torsion dystonia in a child. Acta Paediatr 85(5): 635–637
53. Penn RD (1988) Intrathecal baclofen for severe spasticity. Ann N Y Acad Sci 531: 157–166
54. Penn RD (1992) Intrathecal baclofen for spasticity of spinal origin: seven years of experience. J Neurosurg 77(2): 236–240
55. Penn RD, Kroin JS (1984) Intrathecal baclofen alleviates spinal cord spasticity. Lancet 1(8385): 1078
56. Penn RD, Paice JA (2000) Adverse effects associated with the intrathecal administration of ziconotide. Pain 85(1–2): 291–296
57. Penn RD, Savoy SM, Corcos D, Latash M, Gottlieb G, Parke B, Kroin JS (1989) Intrathecal baclofen for severe spinal spasticity. N Engl J Med 320(23): 1517–1521
58. Pert CB, Snyder SH (1973) Opiate receptor: demonstration in nervous tissue. Science 179: 1011
59. Price GW, Wilkin GP, Turnbull MJ, Bowery NG (1984) Are Baclofen sensitive GABA-B receptors present on primary afferent terminals of the spinal cord? Nature Jan 307(5946): 71–74

60. Rainov NG, Heidecke V, Burkert W (2001) Long-term intrathecal infusion of drug combinations for chronic back and leg pain. J Pain Symptom Manage 22(4): 862–871
61. Rifici C, Kofler M, Kronenberg M, Kofler A, Bramanti P, Saltuari L (1994) Intrathecal baclofen application in patients with supraspinal spasticity secondary to severe traumatic brain injury. Funct Neurol 9(1): 29–34
62. Saissy JM, Demaziere J, Vitris M, Seck M, Marcoux L, Gaye M, Ndiaye M (1992) Treatment of severe tetanus by intrathecal injections of baclofen without artificial ventilation. Intensive Care Med 18(4): 241–244
63. Sampson FC, Hayward A, Evans G, Morton R, Collett B (2002) Functional benefits and cost/benefit analysis of continuous intrathecal baclofen infusion for the management of severe spasticity. J Neurosurg 96(6): 1052–1057
64. Sator-Katzenschlager S, Deusch E, Maier P, Spacek A, Kress HG (2001) The long-term antinociceptive effect of intrathecal S(+)-ketamine in a patient with established morphine tolerance. Anesth Analg 93(4): 1032–1034
65. Taira T, Kawamura H, Tanikawa T, Kawabatake H, Iseki H, Ueda A, Takakura K (1995) A new approach to the control of central deafferentation pains – pinal intrathecal baclofen. Acta Neurochir (Wien) [Suppl] 64: 136–138
66. Thomas JE, Rosenwasser RH, Armonda RA, Harrop J, Mitchell W, Galaria I (1999) Safety of intrathecal sodium nitroprusside for the treatment and prevention of refractory cerebral vasospasm and ischemia in humans. Stroke 30(7): 1409–1416
67. van Hilten BJ, van de Beek WJ, Hoff JI, Voormolen JH, Delhaas EM (2000) Intrathecal baclofen for the treatment of dystonia in patients with reflex sympathetic dystrophy. N Engl J Med 343(9): 625–630
68. Van Schaeybroeck P, Nuttin B, Lagae L, Schrijvers E, Borghgraef C, Feys (2000) Intrathecal baclofen for intractable cerebral spasticity: a prospective placebo-controlled, double-blind study. Neurosurgery 46(3)
69. Walker RH, Danisi FO, Swope DM, Goodman RR, Germano IM, Brin MF (2000) Intrathecal baclofen for dystonia: benefits and complications during six years of experience. Mov Disord 15(6): 1242–1247
70. Wang JK, Nauss LA, Thomas JE (1979) Pain relief by intrathecally applied morphine in man. Anesthesiology 50(2): 149–151
71. Zuniga RE, Perera S, Abram SE (2002) Intrathecal baclofen: a useful agent in the treatment of well-established complex regional pain syndrome. Reg Anesth Pain Med 27(1): 90–93
72. Zuniga RE, Schlicht CR, Abram SE (2000) Intrathecal baclofen is analgesic in patients with chronic pain. Anesthesiology 92(3): 876–880

Correspondence: Michael S. Turner, M.D., Pediatric Neurosurgery, Indianapolis Neurosurgical Group, Inc., Indianapolis, IN, U.S.A. E-mail: TurnerLah@aol.com

Acta Neurochir (2003) [Suppl] 87: 37–38

Clinical application of drug pump for spasticity, pain, and restorative neurosurgery: other clinical applications of intrathecal baclofen

T. Taira and **T. Hori**

Department of Neurosurgery, Neurological Institute, Tokyo Women's Medical University, Tokyo, Japan

Summary

Intrathecal baclofen has been successfully used for control of severe spasticity. Baclofen, an agonist of GABA-B receptor, has other potential effects on pain and recovery from coma. Sporadic episodes of dramatic recovery from persistent vegetative state are reported after intrathecal administration of baclofen. There are also reports on the use of baclofen for neuropathic pain including post-stroke central pain syndrome. Baclofen is also used for control of dystonia due to cerebral palsy or reflex sympathetic dystrophy. On the other hand, epidural spinal cord stimulation has been used for pain, spasticity, dystonia, or attempt to improve deteriorated consciousness, though the effects seem variable and modest. Similarity between baclofen and spinal cord stimulation is interesting in that both involves in spinal GABAergic system. The GABAergic system in the spinal cord plays a pivotal role in various clinical effects of these procedures.

Keywords: Baclofen; pain; dystonia.

Pain management

Baclofen has been used for relief of spasticity for many years. Its analgesic effect, however, is not widely known. In clinical studies, intrathecal baclofen, of course, relieves muscle spasm pain, which is generally believed secondary to relief of spasticity. However, there have been some clinical reports concerning pain relief with intrathecal baclofen. Herman *et al.* [3] reported that central pain caused by spinal lesions is successfully controlled with lumbar intrathecal baclofen and obviously this is not the secondary effect. In their report, a patient even with a C3 lesion experienced relief of pain in the leg. Recently Mertens *et al.* reported that neurogenic pain was relieved during chronic intrathecal infusion of baclofen for treatment of spasticity. We also have reported that intrathecal baclofen effectively suppresses poststroke central pain [5]. Such baclofen analgesia as in patients with central pain of spinal origin can be explained by suppression of the abnormal neuronal activities in the spinal posterior horn. Baclofen analgesia is not mediated through the endogenous opiate system. The neural structures rostral to the medulla and caudal to the midbrain are necessary for the analgesic effect of baclofen. These findings suggest that there is an ascending pain control system from the spinal cord to the pons that is not mediated by the opiate system. Because baclofen acts on GABA-B receptor sites that are present in high concentration in the spinal dorsal horn, GABA may be the mediator of this pain control system. It has been reported that GABA is released by electrical spinal cord stimulation [2, 4], which technique has been clinically used for pain relief for long time. This further supports the importance of the GABAergic system in pain mechanism [8].

Persistent vegetative state

Intrathecal baclofen has been indicated to spasticity of cerebral origin. Some patients with spasticity due to traumatic brain injury or cerebral hypoxia are in so-called chronic vegetative state, and, in such patients, baclofen treatment is generally performed for ease of nursing care. We have experienced dramatic recovery of consciousness after repeated intrathecal injections of baclofen in traumatic brain injured patients [7]. The patients all belonged to the category of "diffuse axonal injury". There are some communications [Becker 1996, Meythaler 1996] from others that they also experienced dramatic recovery of consciousness after intrathecal baclofen. It is known that, in cerebral palsy children, selective dorsal rhizotomy and resultant relief of leg spasticity may show subsequent positive effect on higher brain functions. The effect of baclofen on persistent vegetative state may be secondary, but the effects in some limited cases are so dramatic that we have to consider the primary role of baclofen on disturbance of consciousness. It is known that baclofen improves conduction in demyelinated axons and therefore intrathecal baclofen may accelerate the repair of diffuse axonal injury. Spinal cord stimulation has been used in the hope of recovery from persistent vegetative state, and in some cases, it is really effective. Thus, spinal cord stimulation and baclofen are also similar in terms of recovery from persistent vegetative state.

Dystonia

Spinal cord stimulation used to be reported as effective treatment of dystonia, though the results were not always uniform [10]. Intrathecal baclofen has been introduced for the treatment of generalized dystonia due to cerebral palsy or of unknown etiology, and the results seem promising [1]. It also opened a new therapeutic option for dystonia and pain in reflex sympathetic dystrophy that is refractory to most treatment [9]. Spinal cord stimulation is regarded as a choice of surgical treatment of reflex sympathetic dystrophy.

Cerebral blood flow

Spinal cord stimulation increases cerebral blood flow through unknown mechanisms not related with increased sensory input. The stimulation has been tried in ischemic stroke or vasospasm after subarachnoid hemorrhage. To our knowledge, there is no report on cerebral blood flow and baclofen in clinical setting, while we have experimental data on cerebral blood flow that increased following intrathecal administration of baclofen.

Conclusions

It appears that baclofen and spinal cord stimulation resemble to each other, though the degree of effect on various neurological conditions differs. Both at least involve the spinal GABAergic system. From the facts we presented in this paper, we speculate that the spinal GABAergic system plays an important pivotal role in various clinical effects of baclofen and spinal cord stimulation.

References

1. Albright AL, Barry MJ, Shafton DH, Ferson SS (2001) Intrathecal baclofen for generalized dystonia. Dev Med Child Neurol 43: 652–657
2. Cui JG *et al* (1996) Effects of spinal cord stimulation on touch-evoked allodynia involve GABAergic mechanisms. Pain 66: 287–295
3. Herman RM, D'Luzansky SD, Ippolito R (1992) Intrathecal baclofen suppresses central pain in patients with spinal lesions. Clin J Pain 8: 338–345
4. Stiller CO *et al* (1996) Release of GABA in the dorsal horn and suppression of tactile allodynia by spinal cord stimulation in neuropathic rats. Neurosurgery 39: 367–375
5. Taira T, Kawamura H, Tanikawa T, Iseki H, Takakura K (1994) Spinal intrathecal baclofen suppresses central pain after a stroke. J Neurol Neurosurg Psychiatr 57: 381–386
6. Taira T, Kawamura H, Tanikawa T *et al* (1995) A new approach to the control of central deafferentation pain. Acta Neurochir (Wien) 64: 136–138
7. Taira T, Kawamura H, Tanikawa T *et al* (1997) Dramatic recovery of consciousness after spinal intrathecal administration of baclofen in patients with severe head injury. Treatment Coma 5: 127–134
8. Taira T, Hori T (2000) The contemporary role of neurosurgeons in the management of intractable pain. In: Yanagida H *et al* (eds) Management of pain, a world perspective. Monduzzi, pp 201–204
9. van Hilten BJ, van de Beek WJ, Hoff JI, Voormolen JH, Delhaas EM (2000) Intrathecal baclofen for the treatment of dystonia in patients with reflex sympathetic dystrophy. N Engl J Med 343: 625–630
10. Waltz JM, Davis JA (1983) Cervical cord stimulation in the treatment of athetosis and dystonia. Adv Neurol 37: 225–237

Correspondence: Takaomi Taira, M.D., Ph.D., Department of Neurosurgery, Neurological Institute, Tokyo Women's Medical University, 8-1 Kawada-cho, Shinjuku-ku, Tokyo 162-8666, Japan. E-mail: ttaira@nij.twmu.ac.jp

Acta Neurochir (2003) [Suppl] 87: 39–40

Intrathecal baclofen therapy; patient selection & team approach

A. Nemoto

Department of medical informatics, Yokohama City University Hospital, Japan

Summary

Intrathecal baclofen therapy (ITB therapy) is useful for severe spasticity. However, the Ministry of health, labour and welfare has not allowed the therapy in Japan. A clinical trial of intrathecal baclofen therapy is currently under way. In this paper the situation in Japan with regard to ITB therapy is described and information for physicians in Japan given from our experience. Team approach is important for ITB therapy. Special attention should be paid to patient selection.

Keywords: Intrathecal baclofen therapy; clinical trial; team approach.

Introduction

Intrathecal baclofen therapy (ITB therapy) reduces spasticity but it does not have any effects in paralysis. There are some side effects, such as muscle weakness which means there are some changes in the ADL and APDL of patients. For a good therapy it is important to set a proper goal. Now in Japan, a clinical trial of ITB therapy is going on. It is reported on the situation in Japan and some advice for Japanese doctors is given. Also addressed are patient selection and team approach.

Patient selection

For the ITB therapy patient selection is one of the most important points. Diagnosis, age, body size, severity of spasticity, type and severity of paralysis, ability of communication, ADL, APDL, QOL (including pain, participation, job and so on), should be considered as well as other therapeutic treatments, etc. [5].

At the present indication of ITB therapy is severe spasticity [1, 2, 9]. Typical diagnoses are listed in Table 1. A typical, good candidate to be included has severe spasticity, complete palsy, severe pain in relation to spasticity, difficulties in ADL and no use of spasticity in ADL (i.e. transfer, standing, rolling, seating and so on).

In Japan a clinical trial is being performed with the purpose of gaining allowance of the government for this therapy. Therefore only good candidates should be chosen. However we must not forget the many patients who suffer from severe spasticity in our country and that in the USA and in Europe, targets of ITB become wider and wider [3, 6–8, 10]. For a clinical trial it is important to have good results. Only if this therapy is accepted in our country, we will be able to treat many patients.

As stated before, for the clinical trial the best candidate is a patient who has severe spasticity, complete palsy, severe pain in relation to spasticity, difficulties in ADL and no use of spasticity in ADL. We must be careful with those patients who have severe spasticity with severe pain and complaints, communicative disorder, incomplete palsy, and use of spasticity in ADL.

Table 1. *Indication of ITB*

Spinal origin
– Spinal cord injury
– SCD (familial spastic paraplegia)
OPLL
– Stenosis of spinal canal
– Vascular disorder in spine
– MS
Brain origin
– Cerebral palsy
– Brain injury

Team approach

The necessity for team approach has to be pointed out as I am a physiatrist. In ITB therapy, many specialists are involved: the neurosurgeon who undertakes the operation, the physiatrist who knows the patient's ADL, the neurologist who is familiar with the disease, the orthopedic surgeon who performs other, additional therapies, the physical and occupational therapists who know the patient well and carry out the therapy after the operation, and the medical case worker who takes into account the economical problem, the nurses who take care of a patient after an operation, not to forget the family and the patient himself who have to be informed about the therapy. Most of the patients have a combination therapy such as orthopedic operation and rehabilitation therapies. Doctors and members of a team have to decide on the sequence and the duration of therapies. In case of emergency, such as sudden change of spasticity [4], a neurosurgeon should be available at any time. In order to realize the trouble as soon as possible, physiatrists and therapists are demanded to distinguish even the slightest change. So in many situations this therapy requires a team approach.

It appears that a neurosurgeon or a physiatrist is suited as a team leader, because the neurosurgeon has responsibility for an operation and the physiatrist can judge best on the ADL of a patient. Whoever is the team leader, to have a good contact is most important.

Conclusion

We have to bear in mind the patient's ADL and APDL. Good results will be linked to the best combination of therapies. To get a good result, team approach is very important. ITB is a dream therapy for spasticity, but not an almighty therapy. We have to be aware of the risks and its limitations. ITB therapy will become a boon to the Japanese patients who suffer from severe spasticity.

References

1. Albright AL, Barron WB, Fasick MP, Polinko P, Janosky J (1993) Continuous intrathecal baclofen infusion for spasticity of cerebral origin. JAMA 24; 270: 2475–2477
2. Albright AL (1996) Intrathecal baclofen in cerebral palsy movement disorders. J Child Neurol 11: S29–35
3. Albright AL, Barry MJ, Shafton DH, Ferson SS (2001) Intrathecal baclofen for generalized dystonia. Dev Med Child Neurol 43: 652–657
4. Coffey RJ, Edgar TS, Francisco GE, Graziani V, Meythaler JM, Ridgely PM, Sadiq SA, Turner MS (2002) Abrupt withdrawal from intrathecal baclofen: recognition and management of a potentially life-threatening syndrome. Arch Phys Med Rehabil 83: 735–741
5. Gormley ME Jr (1999) Management of spasticity in children: part 2: oral medications and intrathecal baclofen. J Head Trauma Rehabil 14: 207–209
6. Meythaler JM, Guin-Renfroe S, Law C, Grabb P, Hadley MN (2001) Continuously infused intrathecal baclofen over 12 months for spastic hypertonia in adolescents and adults with cerebral palsy. Arch Phys Med Rehabil 82: 155–161
7. Meythaler JM, Guin-Renfroe S, Brunner RC, Hadley MN (2001) Intrathecal baclofen for spastic hypertonia from stroke. Stroke 32: 2099–2109
8. Nance P, Meythaler J (1999) Intrathecal drug therapy. Phys Med Rehabil Clin N Am 10: 385–401
9. Penn RD (1992) Intrathecal baclofen for spasticity of spinal origin: seven years of experience. J Neurosurg 77: 236–240
10. Taira T, Kawamura H, Tanikawa T, Kawabatake H, Iseki H, Ueda A, Takakura K (1995) New approach to the control of central deafferentation pain – spinal intrathecal baclofen. Acta Neurochir (Wien) [Suppl] 64: 136–138

Correspondence: Akinobu Nemoto, M.D., Department of medical informatics, Yokohama City University Hospital, 3-9 Fukuura, Kanazawa-ku, Yokohama, 236-0004, Japan. E-mail: anemoto@med.yokohama-cu.ac.jp

Current concepts and strategies of early neurorehabilitation

Acta Neurochir (2003) [Suppl] 87: 43–47

New development of functional neurorehabilitation in neurosurgery

K. R. H. von Wild

Medical Faculty of the Westphälische Wilhelms University, Neurosurgical Clinic with Department for Early Rehabilitation of the Clemenshospital, Münster, Germany

Summary

Objectives. Today, increasingly more patients with severe brain and spinal cord lesions mainly secondary to accidents, violence, stroke, and tumours survive their injuries, in many cases, however, suffering from severe functional impairments of functioning as described by the WHO-ICF criteria. New developments of functional neurorehabilitation in neurosurgery could significantly improve the patients' quality of life (QoL) in terms of brain and body functioning and certain health-related components of well-being (such as social activities and leisure).

Methods. Functional rehabilitation is an original task of neurosurgery from the very outset. Advances in biotechnology regarding both basic research and clinical application have opened up a new and very promising field to restore or compensate impaired or definitively lost organic functions in addition to the conservative rehabilitation methods.

Results. Along with the scientific progress in biotechnology and functional MRI and PET, neurosurgeons have become increasingly interested and actively involved in rehabilitation science and neurosurgical re-engineering of the damaged brain and spinal cord. Some of them have developed new specially designed institutions for early (acute) and subacute neurorehabilitation. Attached to the acute services, neurosurgeons thus become responsible for neurorehabilitation and at the same time for the management of all kinds of complications, which significantly improves the early and late functional outcome. At the same time microelectronics, biotechnology, and genetic engineering are being introduced into the field of neurosurgical rehabilitation in a step-by-step manner.

Discussion. Progress in the fields of microelectronics, computer technology, and genetic engineering along with rehabilitation science is opening up a new field of unknown chances to partially restore lost body functions and to help improve the quality of life of disabled patients in the sense of ICF. Functional neurosurgery plays a major role in neurosurgical rehabilitation, e.g. functional electrostimulation, brain-stem implants, pain and epilepsy control, restoration of locomotion and grasp faculties, and the use of potent substances such as botulinum toxin (Btx). This demands the capacity of time work and the realization of the necessity to draw up a detailed plan for the restoration of impaired functions prior to enacting a neurosurgical intervention in the sense of a complex neurorehabilitation, and consequently to assume the responsibility for the patient's outcome. From the beginning of neurological surgery, the preservation and restoration of impaired brain and spinal-cord functions as an original task for neurosurgeons demand their involvement with issues of functional neurorehabilitation including neurosurgical re-engineering of the damaged brain and spinal cord. In this connection the close and trusting cooperation with the clinical neuropsychologist from the very outset is an indispensible factor.

Keywords: Neurorehabilitation; neurosurgical re-engineering; TBI; SCI; functional neurosurgery; international classification of functioning (WHO-ICF).

Introduction

Today ever more patients with severe brain and spinal-cord lesions, mainly after road accidents, personal violence, and also secondary to severe surgical procedures, survive their injuries, in many cases, however, with severe handicaps. This is achieved thanks to the development of modern surgical and intensive care strategies and improved diagnostic methods in the developed countries [2, 3, 5–7, 9–11, 13, 15, 17, 21, 22, 27, 36, 37]. In the countries of the so-called Third World, by contrast, these preconditions are still generally lacking, not to mention those necessary for an adequate follow-up treatment concept. In the meantime there have been many guidelines for the management of severe head and spinal cord injury that have been issued since the first publication by the American Association of Neurological Surgeons and of the Brain Trauma Foundation in 1995 without mentioning even in one instance the necessity or the possibility respectively of functional neurorehabilitation in the setting of general accident care concepts [22, 32, 37]. One reason for this may be the fact that there has until recently indeed been virtually no scientific evidence for the efficacy of neurorehabilitational measures for the early and the late phase and that neurologists and neurosurgeons have devoted any serious attention to the issues of rehabilitational science only in the very recent past [1, 3, 27, 29].

Neurorehabilitation has an old tradition in Europe and especially in Germany. Our actual ideas on and our concepts of neurorehabilitation of sensory motor are based on the ideas of Otfried Foerster (1873 to 1941) [11] regarding the exercise therapy of peripheral paresis and the physiotherapy of central motor disorders, differentiating between the spastic and paretic components. In view of the way in which voluntary movements are generated in a central paretic limb, Foerster had already stated his view by which reactivation takes place via homolateral motor fields. His recommendation was to use movements of the healthy side to trigger corresponding movements of the paretic side, after which it would be possible to gradually model voluntary movements of everyday activities into such a synergy [15, 22, 24, 26, 28]. Today, with the aid of functional MRI studies, we know that we can improve motor skills only by specifically targeted exercises that may be due to the brain's plasticity [6, 19]. In much the same sense Alexander R. Luria (1902–1977), the famous Russian neuropsychologist, saw the activity of the nervous system as a social organ with a coordinated and adaptable cooperation between various segments of the nervous system in which constant results are always yielded under varying conditions [8]. Recent neuropsychological evaluations based on functional magnetic resonance imaging (fMRI) experiments in persons who suffered a lower limb amputation have revealed that their body image was maintained – the amputated leg was shown to be integrated in the body and was perceived in a manner similar to the period preceding the amputation [8.b]. Last but not least Brunelli and I were able to demonstrate the partial restoration of locomotion in a paraplegic patient with total transection of the spinal cord by connecting the upper motor neuron glutamatergic by several nerve grafts from the intact cortical spinal tract to the preserved peripheral muscular nerves (terminal of the lower cholinergic motor neuron). This still experimental operation was based on Brunelli's intensive experience in primates, although the underlying physiological biochemical process remains still unclear [7.b].

Functional electrical stimulation, neuromodulation, bio-implants, and neuroprosthetics open up a highly important field of functional neurosurgical rehabilitation covering movement disorders, spasticity, epilepsy, pain, and vegetative state [4, 13, 18, 24, 26, 27, 29, 37–39]. So what could be recommended as a new development of functional neurorehabilitation in neurosurgery [33]?

Patients and methods

As the available national epidemiological studies and the publications of the WHO impressively show, injuries of the central (CNS) and of the peripheral nervous systems (PNS) – irrespective of the cause – result in enormous costs to society in the national healthcare systems. On top of these come the virtually incalculable long-term costs resulting from the ongoing impairment of organic and cognitive functions and the long-term impairments and social detriments that arise as a consequence. The endeavours of all those persons involved in the patient's care – i.e. the physicians, physiatrists, and therapists – are thus directed towards one common target, namely to provide adequate therapeutical treatment and the necessary aids to the patient at the soonest possible time as a measure to achieve an improvement of his functions, so that the affected, injured patient can be reintegrated into his family, his social environment, and his school or profession. In prosperous countries, there are a number of special training programmes in the field of neurorehabilitation that have been set up in the meantime, accompanied by practical workshops and scientific symposiums. The focus of the interdisciplinary teamwork is directed at the affected patient and his injuries and/or impairments. In particular in those countries affected by war conflicts, in which there is accordingly a high demand for the care of war-injured and war-disabled patients, neurosurgeons and neurologists are increasingly confronted with tasks in the neurorehabilitation field. This is especially true for the phase of acute rehabilitation, with a focus on the issues of neurotraumatology and stroke. Special task forces and scientific societies with an emphasis on neurologico-neurosurgical rehabilitation have in recent years elaborated recommendations and guidelines for the regional and national requirements, aspects that have also been acknowledged and appreciated by the cost-bearing agencies [8, 16, 22, 25, 30, 34]. The result was the establishment of special departments for early rehabilitation and/or post-acute rehabilitation under the leadership or neurosurgeons and neurologists [30, 34]. In Germany, for example, there is now an extensive network for clinic-based acute rehabilitation of patients with brain and/or spinal-cord injuries. In other countries, institutions of this kind are managed by rehabilitation physicians and physiatrists, and the neurosurgeon sees the patients only in the event of a complication. This system – which, in our opinion, is obsolete – is gradually being replaced by the specialized approach of neurologico-neurosurgical rehabilitation. This will presumably occur parallel to the progresses being made in the biotechnology area and in functional imaging and their introduction into the rehabilitation process. Functional imaging (MRI and PET) and the currently available possibilities opened up by electrophysiological diagnostics now make it possible, for the very first time, to follow voluntary or evoked potentials from the site of their origin to the target organ and thus to cast more light onto the wondrous network of neuronal functions [10, 34, 35]. Up to now, merely a relatively small number of patients in the rich developed countries have been able to draw advantages from this technology.

This is similarly true for the gradual clinical introduction of state-of-the-art biotechnology in the areas of neurology and neurosurgery, where brain-stem-similar implants are employed to enhance hearing [18], and also the implantation of microelectrodes and intelligent computer systems, for example for the restoration of locomotion in paraplegic patients [32]. The role played by functional electrostimulation (FES) in the neurorehabilitation field is of particular relevance, where it is applied in the deeper regions of the brain, the nerves of the upper cervical muscles and the spinal cord and in the cerebral and peripheral nerves and for muscle stimulation [4, 13, 27, 37–39].

When listing the recent advances made in functional rehabilitation, one should not forget to mention the special preventive programmes, for instance the cautioning of people living in the proximity of minefields in war zones, or specially elaborated programmes for the prevention of accidents in road traffic, in the household sphere, and industrial accidents.

One of the major services rendered by neuropsychology is the proof that it is essentially mental and cognitive impairments that determine the later life and well-being of the patients, even if it is not always possible to recognize the type and extent of the impairments at the outset. The neuropsychologist thus belongs to the early rehabilitation team as an equal partner from the very start [1, 2, 8, 16, 22, 25, 30, 34].

While the classic triad of physical therapy, physiotherapy, and vocational therapy constitute the undisputed basis for motor rehabilitation, speech therapy, music therapy [1], and neuropsychological cognitive training [16, 22, 25, 31] – wherever possible with the involvement of the patient's family – are becoming ever more important for the outcome, even if the therapeutic success is in most cases empirically derived and reasoned. This notwithstanding, functional imaging techniques and electrophysiological diagnostics have now yielded initial, yet unequivocal evidence of the benefits provided in particular by the exercising, but also by the relaxing methods [25]. International scientific symposiums and the currently available literature show that we are standing on the threshold of a highly promising development in this area. A great number of suitable scales and scores have been published that document the physical and cognitive function gains achieved during neurorehabilitation or, respectively, to describe any shortcomings [22, 36]. In future quality of life in brain injured (QLOBRI) and the ICF of the WHO will be capable of making a major contribution in the assessment of the quality of life [3].

The ICF definition of final functional outcome:

ICDIH definition: The International Classification of Disabilities, Impairments, and Handicaps (ICDIH), which was first published by the World Health Organization for trial purposes in 1980, was endorsed by the 54th World Health Assembly resolution for international use on May 22nd 2001.

The overall aim of the International Classification of Functioning (ICF) is to provide a unified and standard language and framework for the description of health and health-related states. It defines components of health and some health-related components of well-being (such as education and labour). These domains are described from the perspective of the body, the individual, and society in two basic lists:

1. Body functions and structures; and 2. Activities and participation.

These terms, which replace the formerly used terms "impairment", "disability", and "handicap", have been extended in the scope of the classification to allow positive experiences to be described. The new terms are used with specific meanings that may differ from their everyday use. Functioning is an umbrella term encompassing all body functions, activities, and participations. Disability serves as an umbrella term for impairments of activity and limitations of participation restrictions. Functioning and disability associated with health conditions are classified in ICF.

Together, information and diagnosis plus functioning provides a broader and more meaningful picture of the health of people or populations, which can then be used for decision-making purposes. This is of special interest within the spectrum of neurorehabilitation.

Results

The in some cases widespread introduction of neurologico-neurosurgical early rehabilitation in the Anglo-Saxon countries, where the concept is referred to as acute rehabilitation, marks a decisive achievement for the establishment of an instrument for the avoidance of secondary and tertiary complications, while at the same time purposefully giving support to the restorative, compensatory, and maybe also regenerative rehabilitation process [2, 7, 14, 16, 22]. In this regard, the experiences gained in connection with the wars and conflicts of recent years in the medical care of injured military personnel and members of the civilian population have shown a decisive advantage in favour of rehabilitation (Israel, United Kingdom, Russia, United States). In fact, however, the only aspect new in these structures is the multitude of the areas of therapy involved and their coordination in the interdisciplinary team, specially tailored to the individual requirements of the patient. While internal quality control was initially merely the self-imposed guideline of medical and therapeutic activity using state-of-the-art rehabilitation concepts, today in many countries – also in Germany – there are state-controlled and scientifically verified seals of quality that are issued as documentation of the official approval of rehabilitation facilities that meet the specified standards (German Society for neurological Rehabilitation and the QOLIBRI of EMN and EBIS). Such certificates should in future be standardized worldwide as a guarantor for quality-oriented operations in the neurorehabilitation sector. Only when such a path is followed shall the cost-bearing agencies be ready to provide the necessary funds for rehabilitation and reintegration of the injured patient into his family, an adequate vocation, and society in general.

In this connection, the neurosurgeon bears a high measure of responsibility not only in planning, but also in carrying out his surgical interventions. One example in this regard is plastic-reconstructive surgery of a traumatic and a birth-traumatic arm-plexus paresis. The additional gain in function or a further impairment can be calculated early on when this is borne in mind. As experience has shown, this also applies in connection with the restoration of the arm's and hand's grasping function or the standing and walking function of paraplegic patients.

Discussion

One consequence of the intensive research efforts during the Decade of the Brain on the threshold to the new millennium [2, 3, 26, 28] is that the science of rehabilitation medicine – and in particular the area of functional neurorehabilitation – in neurosurgery and neurology has experienced an unprecedented boom, the like of which in earlier times was the inevitable consequence of armed conflicts in the form of a multitude of war injured and war cripples. Interdisciplinary teams of researchers are now focussing their efforts on reconstructing the physiological principles of restorative processes in the brain and spinal cord and on making them exploitable within the clinical setting. Here old theories and hypotheses on neurorehabilitation – for example those of Foerster and von Lurija – find definitive confirmation in the form of the findings provided by state-of-the-art functional imaging techniques such as magnetic resonance tomography (MRI), positron emission tomography (PET scan), and single photron emission computer tomography (SPECT). Just as in the area of motor rehabilitation [15, 32], then according to the current status of our knowledge exercising methods play an overridingly important tale in neuropsychology, meaning that this type of therapy must be provided to the patient as early as possible [8, 14, 22, 25, 30]. Today as ever, the personal relationships between the therapist, the patient's relatives, and the patient himself form the basis for all rehabilitational endeavours. What is more, major advances have also been yielded by developments in the biotechnology field, with no mean influence being exerted by astronautics medicine, where researchers have investigated early and long-term injuries and the extent to which these can be avoided during weightlessness. This relates to both physical training and sensory stimulation procedures as well as to the use of electrophysiological and neurobiological monitoring methods, which have already been specifically adopted in acute and long-term rehabilitation concepts (United States and Russia, France, Germany, Israel, Austria, and Japan). These techniques are still essentially experimental in their nature and are correspondingly limited to a narrow circle of users. This is also true in the case of surgery for the partial restoration of arm, hand-grasp, and walking functions in tetraplegic and paraplegic patients. This notwithstanding, with the tools of biotechnological re-engineering and the utilization of computer-controlled implants we stand at the beginning of a long path [18, 32], and in the next few years we shall doubtlessly see considerable functional advances being achieved in the motor field, presumably also with regard to consciousness, attentiveness, and cognition.

The experience of the past five years has shown that the neurosurgical societies worldwide have recognized functional neurorehabilitation as a challenge for neurosurgeons in the 21st century. 46 major scientific associations have dispatched official delegates to cooperate and to establish and expand special national and international rehabilitation concepts to the Committee for Neurorehabilitation of the World Federation of Neurosurgical Societies, which was founded just a few years ago in 1997. In the meantime not only has the steadily rising need for functional neurorehabilitation been recognized, but also the neurosurgeons have themselves devoted more and more attention to the elaboration of scientific principles and clinical experience in the area of functional neurorehabilitation 36.b. This is evidenced by the fact that neurorehabilitation issues are now being discussed at national and international scientific symposia of the neurosurgical societies at a par with other specialized topics, and that – following the 2001 assembly in Sydney – the 2005 World Congress in Marrakech will go on to scientifically deal with therapeutical issues and interim results under the inclusion of functional-restorative neurosurgery. In this light, the activities of the committee and in particular of the 2nd International Symposium on Neurosurgical Re-engineering of the Damaged Brain and Spinal Cord organized by Prof. Yoichi Katayama in Tokyo, July 10–12, 2002 can indeed be regarded as a central step forward in the philosophy of modern rehabilitation in neurosurgery with the groups of issues dealt with here and with the national and international influence that this committee imparts.

References

1. Aldridge D, Dembski M (eds) (2002) Music Therapy World Musiktherapie – Diagnostik und Wahrnehmung. Private Universität Witten-Herdecke gGmbH ISBN 3-9808248-0-2
2. Barnes MP (1999) Rehabilitation after traumatic brain injury. Br Med Bull 55(4): 927–943
3. Berger E, Leven F, Pierente N, Bouillon B, Neugebauer E (1999) Quality of Life after traumatic brain injury: A systematic review of the literature. Restorat Neurolo Neurosci 14: 93–102
4. Boon P, Vonck K, Van Welleghem P, D'Havé M, Caremaert J, De Reuck J (2001) Vagus nerve stimulation for epilepsy, clinical efficacy of programmed and magnet stimulation. Acta Neurochir (Wien) [Suppl] 79: 93–98
5. Bouillon B *et al* (1999) The Incidence and outcome of severe brain trauma-design and first results of an epidemiological study in an urban area. Restorat Neurol Neurosci 14: 85–92
6. Boyeson MG, Jones JL (1996) Theoretical mechanisms of brain plasticity and therapeutic implications. In: Horn LJ, Zasler ND (eds) Medical rehabilitation of traumatic brain injur. Hanley and Belfus Inc, Philadelphia, pp 77–102

7. Broos PL, Stappaerts KH, Rommens PM (1998) Polytrauma in patients of 65 and over. Injury patters and outcome. Int Surg 73(2): 119–122
7b. Brunelli GA, Brunelli GR (1996) Experimental Surgery in spinal cord lesions by connecting upper motoneurons directly to peripheral targets. J. Periphural Nerv Syst 1(2): 111–118
8. Christensen AL, Uzzel B (eds) (2000) International Handbook of neuropsychological rehabilitation. Kluver Academic/Plenum Publishers, New York Boston Dordrecht London Moscow
8b. Campos da Paz Jr A, Willadino Braga L *et al* (2002) A preliminary functional brain study on amputees SARAH collection 3. SARAH-Letras press Brasilia, pp 5–14
9. Cope DN, Bryant ED, Sundance P (2001) Health management technology for catastrophic medical conditions. Acta Neurochir (Wien) [Suppl] 79: 61–64
10. Diemath HE, Sommerauer J, von Wild KRH (eds) (1996) Brain protection in severe head injury. W Zuckschwerdt-Verlag, München, Bern Wien New York
11. Foerster O (1936) Übungstherapie. In: Bumke O, Foerster O (eds) Handbuch der Neurologie Bd 8. Springer, Berlin Heidelberg New York Tokyo, pp 316–414
12. Gonzáles-Feria L, von Wild KRH, Diemath HE (eds) (2000) Quality management in head injuries care. Servido Canario de Salud, Santa Cruz de Tenerife
13. Grünwald V, Höfener H, Thon WF, Kuczyk MA, Jonas U (1999) Sacral electrical neuromodulation as an alternative treatment option for lower urinary tract dysfunction. Restor Neurol Neurosci 14: 189–193
14. Hellawell DJ, Taylor RT, Pentland B (1999) Cognitive and psychological outcome following moderate or severe traumatic brain injury. Brain Inj 13(7): 489–504
15. Hesse S, Werner C, Bardeleber A, Brandl-Hesse B (2001) Management of upper and lower limb spasticity in neurorehabilitation. Acta Neurochir (Wien) [Suppl] 79: 117–122
16. Hoffmann B, Düwecke C, von Wild KRH (2001) Neurological and social long-term outcome after early rehabilitation following traumatic brain injury. 5 years report on 240 TBI Patients
17. Horn LJ, Zasler ND (eds) (1996) Medical rehabilitation of traumatic brain injury. Hanely & Belfus, Philadelphia, USA
18. Kuchta J, Behr R, Walger M, Michel O, Klug N (2001) Rehabilitation of hearing and communication functions in patients with NF2. Acta Neurochir (Wien) [Suppl] 79: 109–111
19. Lotze M, Laubis-Hermann U, Topka H, Erb M, Grodd W (1999) Reorganization in the primary motor cortex after spinal cord injury – a functional magnetic resonance (fMRI) study. Restor Neurolog Neurosci 14: 183–187
20. Mazaux JM, Croze P, Quintard B, Rouxel L, Joseph P, Richer A, Debelleix E, Barat M (2001) Satisfaction of life and late psycho-social outcome after severe brain injury: a nine-year follow-up study in aquitaine. Acta Neurochir (Wien) [Suppl] 79: 49–51
21. Murray GD *et al* (1999) The European brain injury consortium survey of head injuries. Acta Neurochir (Wien) 141: 223–236
22. Ortega-Suhrkamp E (2001) Early functional outcome in isolated (TBI) and combined (CTBI) brain injury. Acta Neurochir (Wien) [Suppl] 79: 31–32
23. Platz T, Hesse S, Mauritz K-H (1999) Motor rehabilitation after traumatic brain injury and stroke advances in assessment and therapy. Retor Neurol Neurosc 14: 161–166
24. Pagni C, Canavero S, Bonicalzi V, Nurissoi C (2001) The important role of pain in neurorehabilitation. The neurosurgeon's approach. Acta Neurochir (Wien) [Suppl] 79: 67–74
25. Prigatano GP (1999) Working with interdisciplinary rehabilitation treams. In: Prigatano GP (ed) Principles of neuropsychological rehabilitation. Oxford University Press, pp 228–243
26. Sabel BA (1999) Neurotrauma and plasticity. A conference of the German BMBF-research Initiative. Restor Neurolog Neurosc 14: 209–236
27. Schumacher S, Bross S, Scheepe JR, Alken P, Jünemann K-P (1999) Sacral anterior root stimulation andposterior rhizotomy in spastic neuropathic bladder. Restor Neurolog Neurosci 14: 195–199
28. Stein DG (1998) Brain injury and theories of recovery. In: Goldstein LB (ed) Restorative neurology. Futura Publishing Company, Armong, NY, pp 1–24
29. Vesper J, Klostermann F, Funk Th, Stockhammer F, Brock M (2001) Deep brain stimulation of the Globus Pallidus Internus (GPI) for torsion dystonia – a report of two cases. Acta Neurochir (Wien) [Suppl] 79: 83–88
30. Voss A *et al* (1983) Standards der neurologisch-neurochirurgsichen Frührehabilitation. In: von Wild KRH (ed) Spektrum der Neurorehabilitation. W Zuckschwerdt-Verlag, München Bern Wien New York, pp 112–120
31. von Wild KRH (1999) Are there standards in neurotraumatology? Acta Chir Austriaca 31 [Suppl] 159: 23–27
32. von Wild KRH, Rabischong P, Brunelli G, Benichou M, Krishnan K (2001) Computer added locomotion by implanted electrical stimulation in paraplegic patients (SUAW). Acta Neurochir (Wien) [Suppl] 79: 99–104
33. von Wild KRH (2002) Functional rehabilitation in neurosurgery and neurotraumatology. Acta Neurochir (Wien) [Suppl] 79
34. von Wild KRH (2001) Neurorehabilitation – a challenge for neurosurgeons in the 21st century concepts and visions of the WFNS-Committee on neurosurgical rehabilitation. Acta Neurochir (Wien) [Suppl] 79: 3–10
35. von Wild KRH (ed) (1998) Pathophysiological principles and controversies in neurointensive care. W Zuckschwerdt-Verlag, München Bern Wien New York
36. von Wild KRH (2000) Perioperative Management of severe head injuries in adults. In: Schmiedek HH (ed) Operative neurosurgical techniques, 4th edn, vol 1. W B Saunders Company, Philadelphia, pp 45–60
36b. von Wild K, Truelle JL (2004) Quality of life in brain injured (QOLIBRI) EFNS workshop of EMN, Paris, September 2
37. Yamamoto T, Fukaya T, Hirayama T, Katayama Y, Tsumokawa T (1992) Deep-brain and spinal stimulation therapy in persistent vegetative state: changes in PGD2, PGE2 and monoamine in cerebrospinal fluid. Proceedings of the 3nd Annual Meeting of the Society for Treatment of Coma, pp 51–56
38. Yamamoto T, Katayama Y, Osahima H, Fukaya C, Kamamata T, Tsubokawa T (2001) Deep brain stimulation therapy for a persistent vegetative state. Acta Neurochir (Wien) [Suppl] 79: 79–82
39. Yanagida N, Takahashi M, Kinzuka H, Hatazawa J, Uemura K (1999) Dorsal column stimulation for a vegetative patient with hypoxic encephalopathy – an analysis of cerebral metabolism using PET, vol 8. In: Hori S, Kanno T (eds) Neuron Publishing Co, LTd, Tokyo, Japan, pp 37–42

Correspondence: Prof. Klaus R. H. von Wild, Neurosurgery Medical Faculty of the Westfälische Wilhelms University, Frauenburgstr. 32, D-48155 Münster, Germany. E-mail: vonwild@cereb.com; kvw@neuroscience.de

Acta Neurochir (2003) [Suppl] 87: 49–52

Successful treatment by spinal cord stimulation for gait disturbance in a patient with diffuse axonal injury

Y. Hirata[1], **M. Murakami**[1], and **Y. Ushio**[2]

[1] Department of Neurosurgery, Kumamoto Takumadai Hospital, Kumamoto, Japan
[2] Department of Neurosurgery, Kumamoto University Medical School, Kumamoto, Japan

Summary

The authors present a case of diffuse axonal injury (DAI) treated by cervical spinal cord stimulation (C-SCS) for gait disturbance. The patient had right hemiparesis of moderate degree, mild ataxia, ideational apraxia and gait disturbance, when admitted to our hospital for rehabilitation. He could not walk by himself, nevertheless neurorehabilitation was done for four months. Xenon-CT was examined by C-SCS loading and the changes of regional cerebral blood flow were significantly increased in both hemispheres, especially in the thalamus. C-SCS was performed continuously on condition of 25 Hz, 200 microsec and 0.5 V, daily for a month. Neurological deficits, especially gait disturbance due to ideational apraxia, were gradually improved after initiation of C-SCS, and the patient could walk by himself. We speculate that C-SCS played a role in triggering improvement of gait disturbance at the chronic stage in our case, and SCS may be helpful for neurorehabilitation of focal symptoms after DAI.

Keywords: Spinal cord stimulation; diffuse axonal injury; apraxia; cerebral blood flow.

Introduction

Although a large number of reports are available on spinal cord stimulation (SCS), for pain, spasticity and vegetative states [2, 6, 7], there is no report of SCS for patients with focal neurological deficits. On the other hand, the cerebral blood flow increasing effect of SCS has recently been reported elsewhere, both in an experimental and clinical setting [8, 12–15]. We performed SCS on a patient with gait disturbance due to ideational apraxia in chronic stage following diffuse axonal injury (DAI) and assessed its clinical effect. Also, cerebral blood flow (CBF) was measured by Xenon-CT (Xe-CT) before and during SCS to examine the CBF increasing effect of SCS.

Case report

History

A 24-year-old man had suffered a traffic accident on May 3, 2001 and the patient was admitted to a hospital in Miyazaki City. Neurological examination revealed mild restless confusion and mild right hemiparesis. Glasgow coma scale was 13 points (E4, V4, M5). CT scan showed a thin interhemispheric subdural hemorrhage and a small hemorrhage in the corpus callosum, confirming the diagnosis of DAI. He had gradually improved since onset, however, had suddenly generalized convulsions and was semicomatose 8 days after onset. CT scan showed a thick interhemispheric and cerebral hemorrhage. Consciousness disturbance and pulmonary edema were gradually improved. The patient was admitted to our hospital for neurorehabilitation 3 months after onset on July 30, 2001. Neurological examination revealed right hemiparesis of moderate degree with spasticity, mild ataxia on the right side, gait disturbance, ideational apraxia and mild motor aphasia. Neurorehabilitaiton was applied for four months. Magnetic resonance imaging demonstrated a thin contusional area in the frontal lobe, corpus callosum and in the medial surface of the occipital lobe (Fig. 1). After neurorehabilitation for four months, right hemiparesis and motor aphasia were markedly improved, however, gait disturbance and ideational apraxia remained unchanged. His gait disturbance was very unusual. He could not use a stick for gait, although he knew what the stick was.

Operation

Spinal cord stimulation (SCS) was performed using a Medtronic system (Medtronic Neurological, Minneapolis, MN). A tetrapolar electrode (Pisces-Quad; Medtronic Neurological) was percutaneously inserted under local anesthesia. The electrode was placed on the posterior surface of the spinal cord at C2–C4 in the epidural space. An external impulse generator provided an adjustable range of pulse width, intensity and frequency of stimulations. An SCS "test" voltage level (0.5 V) was selected by provoking mild paresthesia in the upper limbs. Pulse width was 200 microsec, and the rate was 25 Hz. C-SCS was performed for 6 hours every day. The electrode was removed after one month.

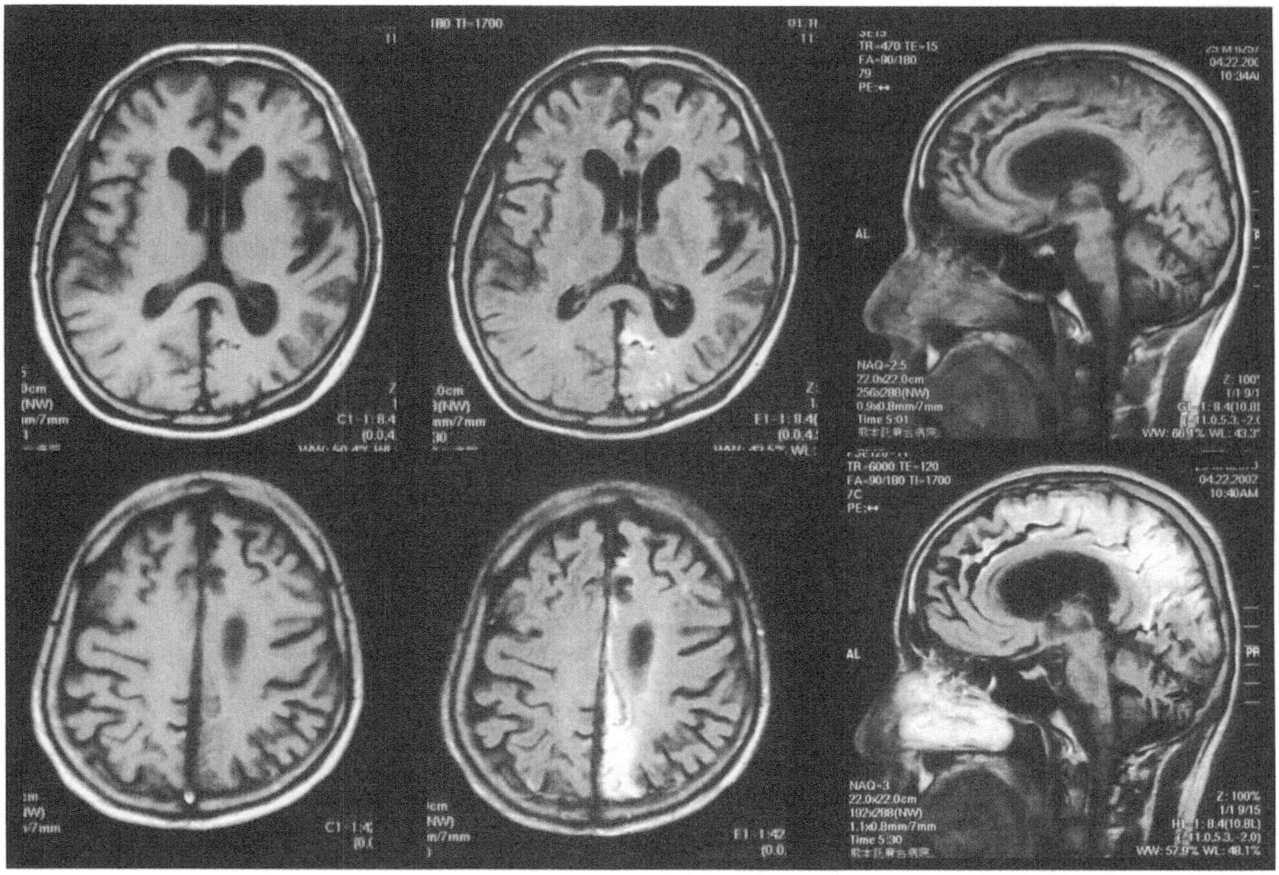

Fig. 1. MRI demonstrated a thin high intensity in the left frontal lobe. Corpus callosum and in the medial surface of the occipital lobe 8 months after onset

Measurement of r-CBF

A Toshiba model X-vision GX CT scanner (Toshiba Medical Co., Tokyo, Japan) was used for Xe-CT studies. A closed-circuit anesthesia apparatus (Xetron AZ-725, Anzai Medical Co., Ltd., Tokyo) was used to administer xenon gas by inhalation. For the Xe-CT study, regions of interest (ROI) were defined in six areas of the hemisphere; the frontal lobe, temporal lobe, occipital lobe, thalamus, basal ganglia and hemisphere. The CBF in the six ROIs was measured in which 30% cold xenon gas (Xenon-Cold, Anzai Medical Co., Ltd) was inhalated for four minutes, the CBF was calculated during the CBF image processing of system (AZ-7000W, Anzai Medical Co. Ltd.). Xe-CTs were performed twice before SCS and after SCS for 30 minutes on day of operation, and after SCS for a month. The changes of r-CBF over 10% of the pre-loading of SCS were defined as significant. Before SCS, the CBF values in every ROI of the left hemisphere were lower than those of the right hemisphere, especially in the left temporo-occipital lobe. After SCS for 30 minutes, the CBF values in every ROI of the right hemisphere were higher than those before SCS. Furthermore, the r-CBF values in the left hemisphere after SCS for 30 minutes were compared with those before SCS, a significant difference was seen in the frontal lobe, temporal lobe and thalamus. After SCS for a month, the r-CBF values in both hemispheres were still increased, especially in the thalamus, as compared to those before (Fig. 2a, b).

Postoperative course

During temporary SCS for a month, the patient could walk with a stick by himself. Ideational apraxia was improved and spasticity had also decreased. After a lead was removed, the neurological symptoms remained improved. He went home three months after temporary SCS.

Discussion

SCS is a useful method for the treatment of pain and ischemic syndromes [1]. Increase in cerebral blood flow (CBF) during SCS has been demonstrated using the inhalated Xenon-133 washout method [6], TCD [12] and, more recently, by functional magnetic resonance imaging [10] and positron emissin tomography [4]. Hosobuchi [7] reported a 34% increase in cerebral blood flow by SCS in patients with chronic pain. Fujii *et al.* [2] reported a CBF-increasing effect of SCS in vegetative patients. Clavo *et al.* [3] found increase of brain tumor oxygen during C-SCS and hypothesized

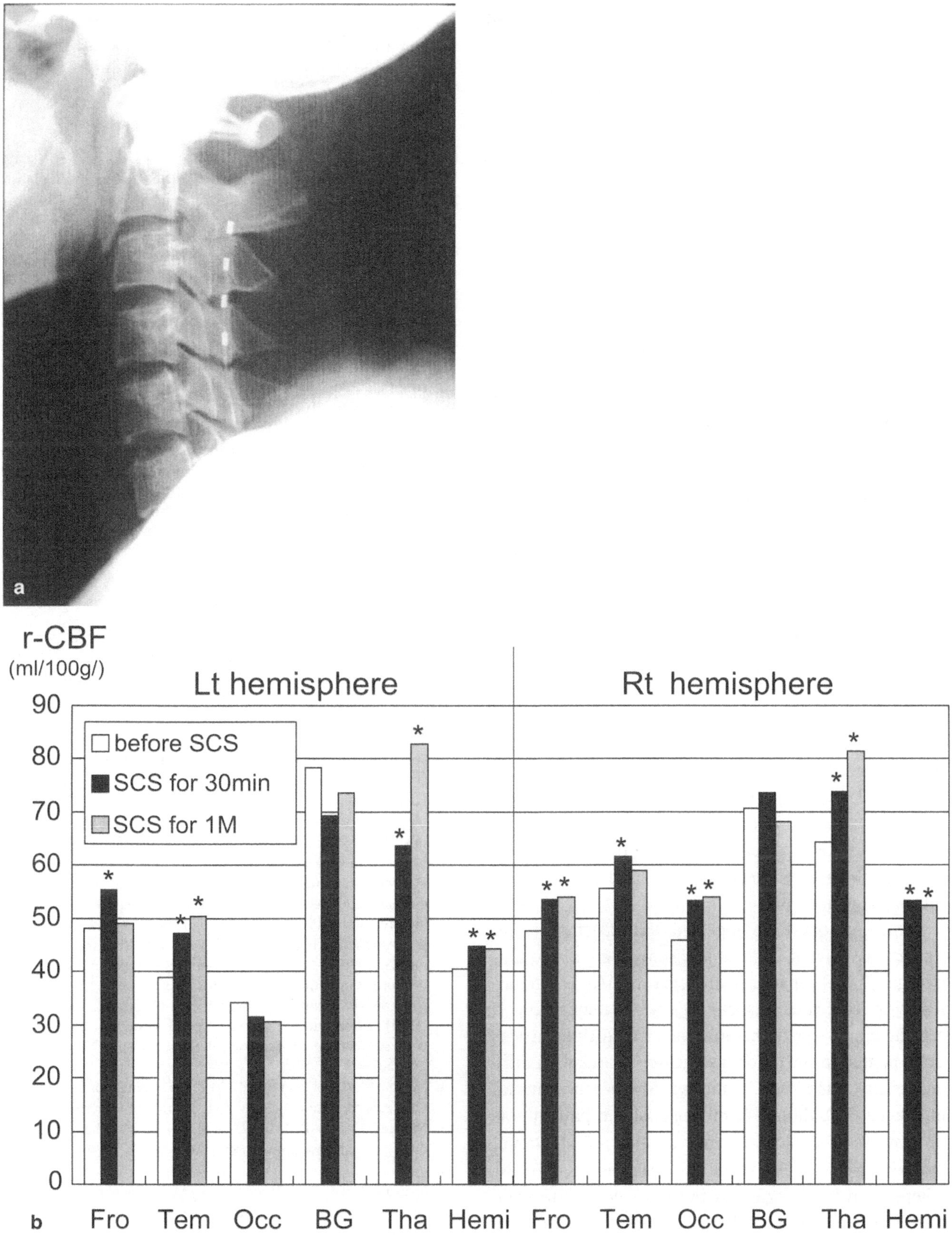

Fig. 2. (a) Cervical X ray film showed a 4 polar electrodes of PISCES placed in the poeterior epidual space of the high cervial spinal region. (b) Measurements of regional cerebral blood flow (*r-CBF*) were demonstrated before SCS, after SCS for 30 min and after SCS for a month. r-CBF values were especially significantly increased in thalamus. * The changes of r-CBF over 10% of the pre-loading of SCS were defined as significant

that SCS can increase regional blood flow in brain tumors. We have reported that high cervical SCS was attempted in five patients with chronic consciousness disturbance of moderate degrees after subarachnoid hemorrhage and that activities, difficulty in eating and spasticy were gradually improved [5]. In the chronic stage of DAI, hypoperfusion has been demonstrated by SPECT despite the abscence of CT evidence for parenchymal change [11]. In our case, r-CBF showed hypoperfusion in both hemispheres on Xe-CT, although a cerebral contusion was demonstrated only in the medical surface of the left hemisphere. The present case showed improved CBF of both hemispheres in response to SCS, even after a long time lag from onset of DAI. No reports are available on SCS for patients like ours, in whom focal neurological symptoms were improved by temporary SCS. We speculate that the gait disturbance and ideational apraxia were improved after increase of r-CBF. The present case was initiated to evaluate the hypothesis that C-SCS can increase r-CBF in DAI. The action mechanism of SCS for increase of r-CBF remains unclear, an effect on the automatic nervous system has been suggested. Isono *et al.* [8] experimentally suggested sympathetic nervous system suppression. Hosobuchi [6] assumed the humoral factors, and Mazzone *et al.* [13] speculated on the activation of brainstem reticular formation since they observed increased CBF in the frontal lobe after SCS in 12 patients with spasm or pain. We also assume increased CBF via thalamo-cortical projection, because the CBF values of thalamus, frontal lobe and temporal lobe were especially increased by C-SCS in our present case.

Some authors have reported that DAI was frequently associated with spastic hemiplegia and quadriplegia [9, 11]. Spastic hemiplegia in DAI is a manifestation of primary shear injury. Matsuzawa [11] reported that spastic hemiplegia of quadriplegia was documented in the chronic stage in 63 cases with difficulty in walking. Gait disturbance was very unusual in our case. Although hemiplegia of the patient was improved, he could not walk because of ideational apraxia. No case like ours has been reported so far where focal neurological symptoms, such as apraxia, were improved by SCS. We speculate that temporary SCS played a role in triggering improvement of gait disturbance in this case. SCS may be helpful for neurorehabilitation of focal symptoms after DAI.

References

1. Broseta J, Barbea J, de Vera JA (1986) Spinal cord stimulation in peripheral arterial disease. A cooperative study. J Neurosurg 64: 71–80
2. Fujii M, Sakamoto K, Yun E, Nakajima K, Kitahara T, Ishihara H, Fujisawa H, Yamashita T, Ito H (1995) The evaluation of spinal cord stimulation (SCS) therapy for patients waith prolonged coma using presurgical Xenon-CT CBF. The Society for Treatment of Coma 4: 187–195
3. Gunderoth M (2002) Increase of brain tumor oxygen during cervical spinal cord stimulation. Report of three cases. J Neurosurg (Spinal) 96: 94–100
4. Hartvast RW, Ter Horst GJ, De Jong BM (1997) Relative changes in regional cerebral blood flow during spinal cord stimulation in patients with refractory angina pectoris. Eng J Neurosci 9: 1178–1183
5. Hirata Y, Ushio Y (1998) Spinal cord stimulation in the treatment of patients with chronic consciousness disturbance of moderate degree after subarachnoid hemorrhage. The Society for Treatment of Coma 7: 43–47
6. Hosobuchi Y (1985) Electrical stimulation of the cervical spinal cord increases cerebral blood flow in humans. Appl Neurophysiol 48: 372–376
7. Hosobuchi Y (1991) Treatment of cerebral ischemia with electrical stimulation of the cervical spinal cord. PACE 14: 122–126
8. Isono M, Kaga A, Fujiki M, Mori T, Hori S (1995) Effect of spinal cord stimulation on cerebral blood flow in cats. Stereotact Funct Neurosurg 64: 40–46
9. Jennet B, Snoek J, Bond MR, Brocks (1981) Disability after servere head injury: observations on the use of the Glasgow coma scale. J Neurol Neurosurg Psychiatry 44: 285–293
10. Kiriakopoulos ET, Tasker RR, Nicosia S (1997) Functional magnetic resonance imaging: a potential tool for the evaluation of spinal cord stimulation. Neurosurgery 41: 501–504
11. Matsuzawa H, Kubo T, Kanazawa I, Kamitani H, Nakamura N (1997) Shearing injuries of parasagittal white matter, corpus callosum and basal ganglia; possible radiological of hemiplegia in diffuse axonal injury. Neurosurg 25(8): 689–694
12. Mazzone P, Rodriguez G, Arrigo A, Nobili F, Pisani R, Rosadini G (1996) Cerebral hemodynamics changes induced by spinal cord stimulation in man. Ital J Neurol Sci 17: 55–57
13. Mazzone P, Pisani R, Nobili F, Gamburo M, Rodriguez G (1995) Assessment of regional cerebral blood flow during spinal cord stimulation in humans. Stereotact Funct Neurosurg 64: 197–201
14. Meglio M, Cioni B, Visocchi M (1991) Spinal cord stimulation and cerebral hemodynamics. Acta Neurochir (Wien) 111: 43–48
15. Meglio M, Cioni B, Visocchi M (1991) Cerebral hemodynamics during spinal cord stimulation. PACE 14: 127–130

Correspondence: Yoshifumi Hirata, M.D., Department of Neurosurgery, Kumamoto Takumadai Hospital, 14-27, 1-Chome, Onoue, Kumamoto, 862-0913, Japan. E-mail: hisyo@po.infobears.ne.jp

Acta Neurochir (2003) [Suppl] 87: 53–55

Functional Electrical Stimulation (FES) for spinal cord injury

I. Okuma[1], **J. Hayashi**[1], **T. Kaito**[2], **M. Funahashi**[2], **S. Kuno**[1], **Y. Kato**[1], **M. Shoda**[1], **J. Nagata**[1], **M. W. Keith**[3], and **T. Kanno**[1]

[1] Department of Neurosurgery, Fujita Health University, Toyoake, Japan
[2] Department of Neurosurgery, TOYOTA Memorial Hospital, Toyota, Japan
[3] Department of Orthopedics, Case Western Reserve University, Ohio, U.S.A.

Summary

Restoration of respiratory motion by stimulation of the phrenic nerve was investigated. Respiratory motion was restored successfully by introducing a breathing pacemaker to a patient with respiratory disturbance due to upper cervical spinal cord injury. Breathing pacemakers are considered to be more similar to physiological conditions compared to mechanical ventilators. Although the system is very expensive, its cost effectiveness may be excellent, provided that it can be used for long hours each day over an extended period. The system is effective in improving patient QOL because it dramatically increases patient mobility. From these findings, it is concluded that breathing pacemakers should be used more frequently in Japan, and that various forms of support are necessary to cope with economic and other concerns.

Keywords: Breathing pacemaker; diaphragm pacing; functional electrical stimulation.

Introduction

Upper cervical spinal cord injuries cause not only limb paralysis but also respiratory function disorder. Affected patients must therefore undergo continuous positive pressure respiratory management using a mechanical ventilator. This limits their behavior considerably, and is a major disruption to rehabilitation and home care. Traditionally, there have been therapeutic attempts to use electrical stimulation as an aid to restore motor function in spinal cord injuries or cerebral stroke. This approach is known as functional electrical stimulation (FES). In recent years, advances in computer technology have enabled precise control of FES, and this is now a clinically useful therapy. Against this background, we introduced a breathing pacemaker to restore diaphragm motion and normal respiration by electrically stimulating the phrenic nerve.

Materials and methods

The subject of this study was a 20-year-old male with C2 cervical spinal cord injury associated with dislocation fracture. Emergency surgery was performed to achieve anterior fusion. After surgery, the patient was lucent but presented limb paralysis and respiratory disturbance as sequelae, requiring respiratory management using a mechanical ventilator. Our attempts to wean the patient from the mechanical ventilator were unsuccessful with no sufficient tidal volume obtained for 18 months.

Three types of breathing pacemakers are available. We selected the pacemaker manufactured by Avery. First, the phrenic nerve was stimulated percutaneously to examine the condition of the phrenic nerve. Specifically, phrenic nerve conduction time was determined and diaphragm constrictions were confirmed by radiographic imaging.

The surgical technique is described below. A 5-cm skin incision was made about 2 cm above each clavicle. The platysma was divided and the lateral margin of the sternocleidomastoid muscle was dis-

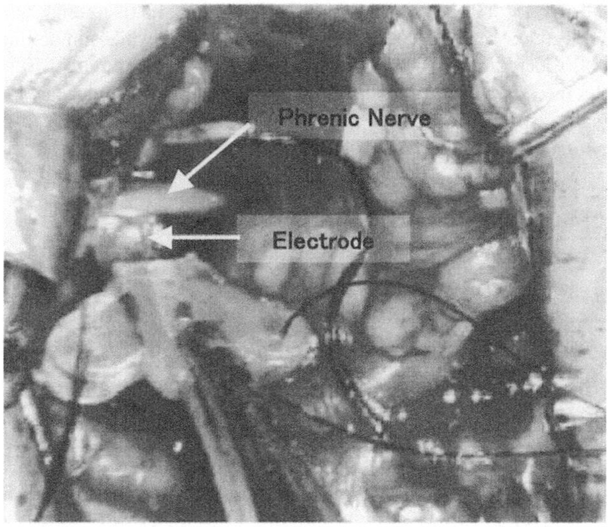

Fig. 1.

Table 1. *Stimulation pulse train*

– Amplitude (V)	
Rt-1.2, Lt-1.1 (0∼10)	
– Frequency (Hz)	
20	
– Pulse interval (msec)	
50 (40∼130)	
– Pulse width (usec)	
150	
– Inspiratory period (sec)	
1.3 (1.2∼1.45)	
– Rate (Breath/min.)	
8 (6∼24)	

sected to the adipose tissue on the anterior face of the anterior scalene muscle. While carefully dissecting the adipose tissue, the position of the phrenic nerve was confirmed by electrical stimulation. Subsequently, the fascia surrounding the phrenic nerve was dissected and a tunnel made under the nerve. During this procedure, no electrical surgical knives etc. were used. An electrode was then set in place. Specifically, the electrode was placed around the nerve horizontally, and fixed by suturing in a manner such that no torsion was produced and no excess tension was exerted on the nerve (Fig. 1). At the same time, an electrical stimulation was applied to confirm constrictions of the diaphragm. A pocket was made subcutaneously in the precordia and a receiver was inserted. All leads were passed beneath the skin and connected to the electrode. Finally, normal operation of the system was confirmed and the wound was closed. This implantation surgery was performed bilaterally in one procedure.

The system was not used until four weeks after surgery to ensure a calm condition around the phrenic nerve, and to alleviate the edema surrounding the wound. High compression and major movement of the neck were also avoided. After the rest period, the breathing pacemaker was switched on, with the threshold and maximum amplitude being first determined by radiographic imaging. At the same time, the diaphragm constriction status was checked on both sides. Initially, the nerve was stimulated at a rate of about 5 minutes per hour, and the stimulation intensity suitable for the patient was determined on the basis of findings regarding the presence or absence of pain at the receiver portion upon stimulation, tidal volume, and subjective symptoms. We determined these factors while stimulating both sides simultaneously. In an attempt to wean the patient from the mechanical ventilator, the length of electrical stimulation was increased each week, with special attention paid to possible diaphragm fatigue due to electrical stimulation.

Results

The likelihood of diaphragm fatigue was reduced by replacing fatigue-prone myoplasm with fatigue-resistant myoplasm while being aware of any reduction in ventilatory volume. One month after stimulation initiation, it became possible to provide 1-hour continuous stimulation three times a day. Although the patient initially complained of a sensation of respiratory disorder during operation of the system, this symptom was ameliorated in about 10 days. At three months, 3-hour continuous stimulation was possible; the breathing pacemaker was operated during daily activities, including food intake, rehabilitation and computer use. At five months, the patient was able to have the system in use for 10 consecutive hours each day, and there was almost no change in ventilatory volume between initiation and completion of stimulation, with sufficient effect obtained at a stimulation intensity of 1.2 V to 1.5 V and a ventilation frequency of 8 times/min (Table 1, Fig. 2). The patient was weaned from the mechanical ventilator full time.

Discussion

Breathing pacemakers have a long history; the world's first breathing pacemaker was used by Glenn *et al.* at Yale University in 1959. To date, more than 1,000 clinical cases have been reported worldwide. Additionally, investigations have been conducted on electrodes for stimulation of the phrenic nerve and attempts have been made to stimulate the intercostal nerve or the abdominal rectus muscle simultaneously to further increase tidal volume.

Indications of breathing pacemakers include central respiratory disturbance due to brainstem or cervical spinal cord injuries, central alveolar hypoventilation syndrome (Ondine's curse), nocturnal hypoventilation syndrome, and uncontrollable hicups (Table 2). It is necessary to stimulate the phrenic nerve percutaneously before surgery to confirm good constrictions of the diaphragm. However, percutaneous stimulation of the phrenic nerve is difficult; excess stimulation intensity causes severe pain at the stimulation site, which in turn is likely to produce false negative test results. Although it seems that satisfactory results are often obtained with intraoperative stimulation of the phrenic nerve in the neck or chest, we believe that diaphragm constrictions and conduction time should always be confirmed. As a general rule, stimulation was conducted bilaterally at one time during operation of the breathing pacemaker, since these conditions are most similar to physiological conditions, since bilateral stimulation ensures a sufficient ventilatory volume even if stimulation intensity is reduced near the threshold, and also since ventilation frequency can be minimized. With these means, we attempted to cope with the major problem of diaphragm fatigue in breathing pacemakers, although there are other factors involved, including diaphragm condition, sur-

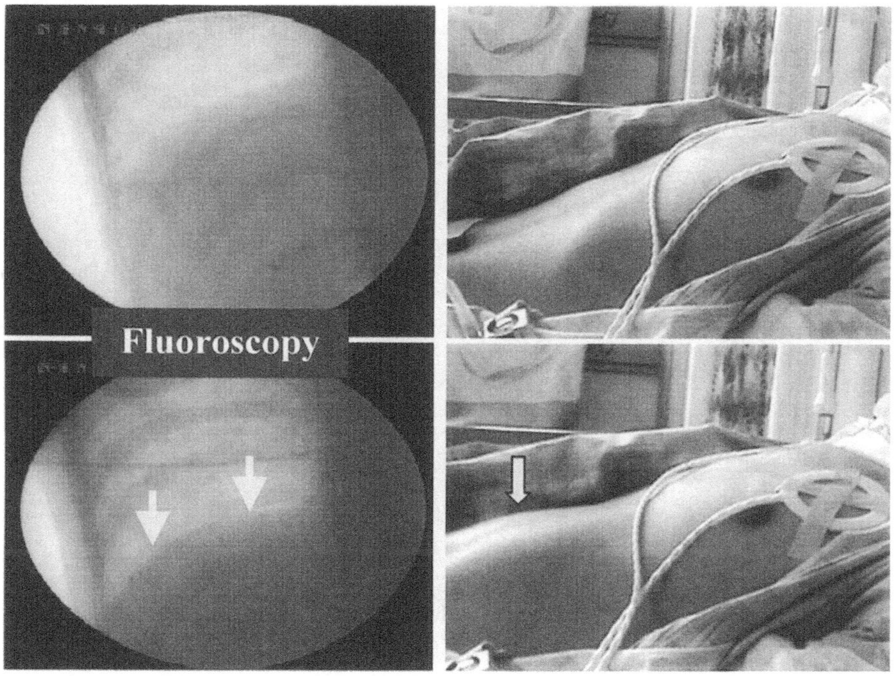

Fig. 2.

Table 2. *Breathing pacemaker indications*

– Above C4 level spinal cord injury (especially C2)
– Congenital central hypoventilation syndrome (ondine curse)

gical procedures, training method, and stimulation method. Currently, after 8 months of electrical stimulation treatment with no reduction in ventilatory volume, the patient is able to receive three courses per day of 5-hour electrical stimulation at 1-hour rest intervals, and is being trained to use the breathing pacemaker for extended periods. This patient was paced full time. Regarding body position-related changes in ventilatory volume, there has been no problem in the present patient. The breathing pacemaker is operated while the patient is in a sitting position; we are ready to consider abdominal immobilization with sandbags etc. and auxiliary stimulation of the abdominal rectus muscle or the intercostal muscle whenever necessary. In regard to the patient, he feels a sensation of respiratory disorder for a while when placed on the mechanical ventilator following use of the breathing pacemaker, complaining of a choking feeling despite the absence of problematic data on ventilatory volume or blood gas. Attributable to the low stimulation intensity used, diaphragm fatigue or pain at the stimulation site has not been observed (these symptoms are often problematic in functional electrical stimulation). Because respiration is essential to living, even a very slight sensation of respiratory disorder can be a major stress. Such problems should be investigated with an increased number of subjects.

References

1. Baue AE (1992) Glenn's thoracic and cardiovascular surgery, 6th edn, vol 1. Prentice-Hall International Inc
2. Exner G (2000) Functional electrical stimulation in paralyzed respiratory muscles: International Workshop in Hamburg, Germany. Neuromodulation 3(4): 211–217
3. Keith MW (1989) Implantable functional neuromuscular stimulation in the tetraplegic hand. J Hand Surg 14A: 524–530
4. Okuma I (2000) Functional electrical stimulation. In: Takakura K (ed) The state of the art of neurosurgery, 2nd edn
5. Talonen PP (1990) Neurophysiological and technical considerations for the design of an implantable phrenic nerve stimulator. Med & Biological Engineering Computing 28 January: 31–37

Correspondence: Isao Okuma, M.D., Department of Neurosurgery, Fujita Health University, 1-98 Dengakugakubo, Kutsukake, Toyoake, Aichi, Japan. E-mail: neuron@fujita-hu.ac.jp

Functional imaging in neurosurgery and neurorehabilitation

Acta Neurochir (2003) [Suppl] 87: 59–62

BOLD functional MRI may overlook activation areas in the damaged brain

K. Sakatani, Y. Murata, C. Fukaya, T. Yamamoto, and **Y. Katayama**

Department of Neurosurgery, Nihon University School of Medicine, Tokyo, Japan

Summary

Clinical applications of blood-oxygenation-level-dependent contrast functional MRI (BOLD-fMRI) have been rapidly moving toward routine non-invasive cortical mapping in the patients with brain disorders. However, it is not yet clear whether the damaged brain shows same cerebral blood oxygenation (CBO) changes during neuronal activation as those in the normal adult. We compared the activation mapping obtained by BOLD-fMRI and the evoked-CBO changes measured by near infrared spectroscopy (NIRS) in normal adults (6 cases) and patients with damaged brain (6 cases of cerebral ischemia and 10 cases of brain tumors in or adjacent to the motor cortex). BOLD-fMRI demonstrated robust activation areas in the primary sensorimotor cortex (PSMC) during contralateral hand grasping tasks in all of the normal adults; however, in the cerebral ischemia (6 cases) and the brain tumors (2 cases), BOLD-fMRI demonstrated only limited activation areas in the PSMC on the lesion side during the task. NIRS demonstrated an increase of focal concentration of oxyhemoglobin and total hemoglobin at the PSMC during the task in all of the normal adults and the patients, indicating the presence of rCBF increase in response to neuronal activation. A focal concentration of deoxyhemoglobin decreased during the task in the normal adults, however, in the patients that showed limited activation areas by BOLD-fMRI, deoxyhemoglobin concentrations increased during the entire course of the task. In summary, the evoked-CBO changes occurring in the damaged brain differed from those in the normal brain. This indicates that BOLD-fMRI may overlook activation areas in the damaged brain.

Keywords: NIRS; BOLD; cerebral blood oxygenation; motor cortex.

Introduction

Clinical applications of blood-oxygenation-level-dependent contrast functional MRI (BOLD-fMRI) have been rapidly moving toward routine non-invasive cortical mapping in patients with various brain disorders, including damaged brain [1, 2].

PET activation studies on human visual or somatosensory functions demonstrated that rCBF is greatly increased by focal increases in neuronal activity; whereas, the cerebral metabolic rate for O_2 ($CMRO_2$) increases only slightly [3, 4]. At the activating cortical area, the rCBF increases by 50% while $CMRO_2$ increases by no more than 5%. This uncoupling of rCBF and $CMRO_2$ during neuronal activity leads to a decrease in the extracted fraction of available O_2, and therefore, Deoxy-Hb concentrations decrease in the cerebral vessels. The decreased concentration of Deoxy-Hb during neuronal activity increases BOLD-fMRI signals because Deoxy-Hb is paramagnetic [5].

However, it is not yet clear whether the damaged brain shows same cerebral blood oxygenation (CBO) changes during neuronal activation as those in the normal adult. Neuronal activation studies employing near infrared spectroscopy (NIRS) have suggested that the evoked-CBO changes in brain disorders differ from those of normal adults [6]. To address this issue, we compared the activation mapping obtained by BOLD-fMRI and the evoked-CBO changes measured by near infrared spectroscopy (NIRS) in normal adults and patients with damaged brain.

Patients and methods

We studied six normal adults (28–68 y), six cerebral ischemic patients (42–68 y) and 10 cases (24–65 y) of glioma in or adjacent to the primary sensorimotor cortex (PSMC). In the cerebral ischemic patients, three cases had internal carotid artery occlusion, two cases had middle cerebral artery occlusion, and one case had middle cerebral artery stenosis. CT or MRI was undertaken in each patient to obtain morphological information prior to the examination; one case showed cerebral infarction, while five cases revealed no lesion. Single photon emission CT (SPECT; ECD-RVR method) revealed a low baseline rCBF (27 ± 4.1 ml/g/min) and poor vascular response to acetazolamide in the PSMC on the lesion side of the ischemic patients. In the glioma patients, activation in the PSMC was confirmed by cortical mapping or cortico-spinal motor evoked potential during surgery in all patients. Informed consent to participate in the present study was obtained from each subject.

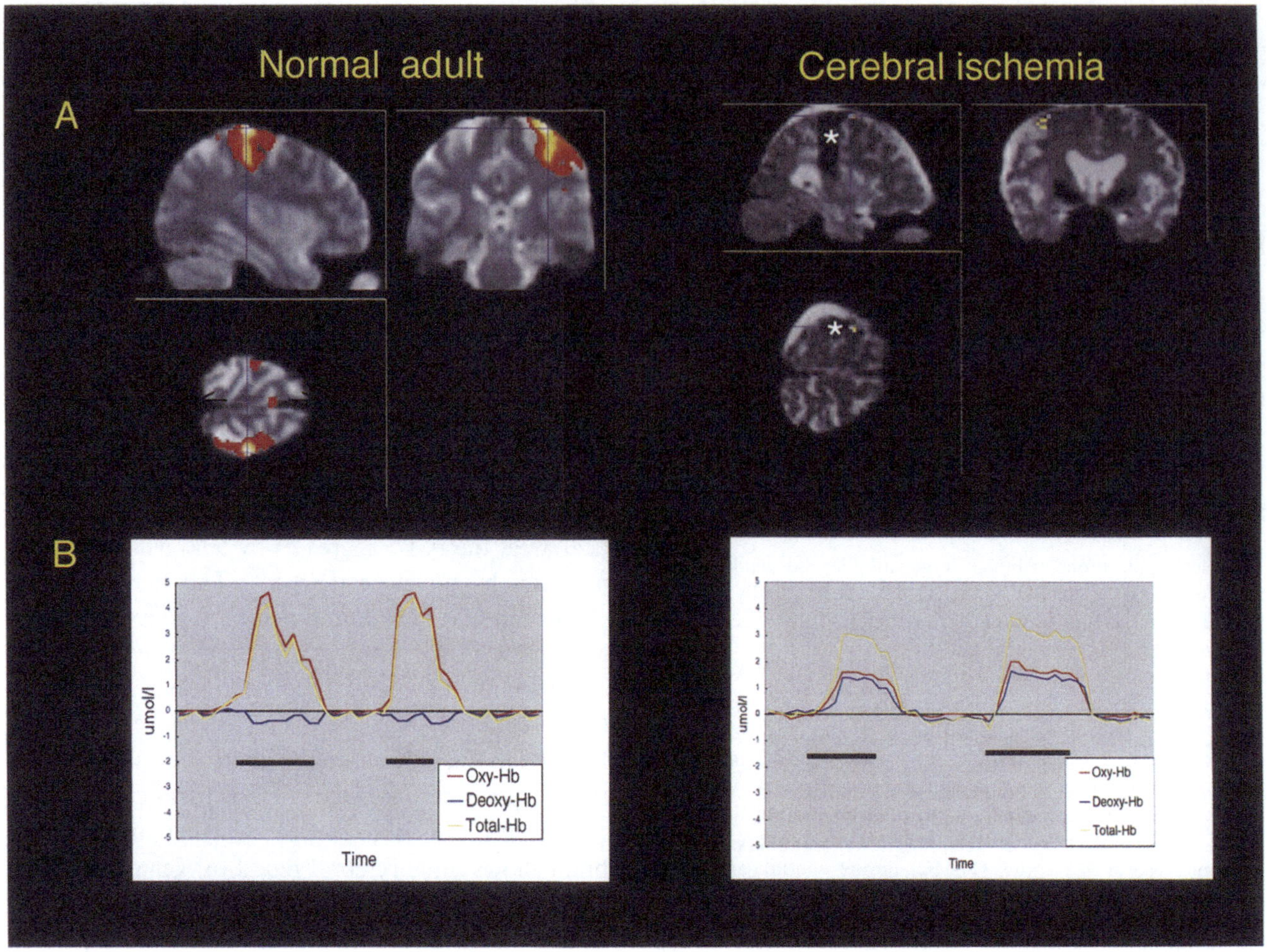

Fig. 1. (A) Threshold activated areas superimposed onto T_2 images of the normal (right motor task; left) and the patient (left motor task; right). The asterisk in the right indicates a lesion of infarction. (B) CBO changes in the PSMC of the normal (left) and the cerebral ischemic patient (68 y, right internal carotid artery occlusion, 12 months after onset; right) during contralateral motor tasks measured by NIRS. Deoxy-Hb decreased in the normal, but increased in the patient. The ordinates indicate concentration changes of Oxy-Hb, Deoxy-Hb, and Total-Hb. Horizontal thick bars indicate the period of the task

Hand grasping was performed as a motor task, and all patients could perform the task similarly to the normals at the time of examination. We employed a NIRO-300 monitor (Hamamatsu Photonics K.K., Japan) to measure the concentration changes of Deoxy-Hb, Oxy-Hb and Total-Hb in the PSMC contralateral to the task performance. The probes were placed at a distance of 3 or 4 cm on the head over the PSMC, so that the axis of the probes could be aimed to superimpose on the central sulcus; 3 cm posterior to the bregma, and 30 degrees outside from the median line of the head [7]. After carrying out the initial setting by this method, the position of the probes was adjusted so that the maximum responses of Oxy-Hb and Total-Hb were obtained during the task performance. In five normal subjects, the location of the probes was identified by MRI utilizing vitamin E capsules.

BOLD-fMRI signals were measured by 1.5 T MRI (Symphony, Siemens, Germany) using an echo-planar technique; TE 50 msec, TR 4 s, slice thickness 3 mm, matrix size 40×40, FOV 192×192 mm. 120 frames of 40 axial slices (acquisition time of one frame 4 sec) through the PSMC were acquired during repeated motor task (40 sec) and resting periods (40 sec); this task-rest cycle was repeated 6 times. Activation maps were calculated by Statistical Parametric Mapping (SPM; Z-score > 1.5).

Results

In the normal subjects, BOLD-fMRI demonstrated robust activation areas in the PSMC contralateral to the performing hand (Fig. 1A left). NIRS demonstrated a decrease of Deoxy-Hb associated with increases of Oxy-Hb and Total-Hb in the PSMC contralateral to the task performance (Fig. 1B left).

However, in the cerebral ischemia (6 cases), BOLD-fMRI demonstrated only limited activation areas in the PSMC on the lesion side during the task (Fig. 1A right). NIRS demonstrated an increase of focal concentration of Oxy-Hb and Total-Hb at the PSMC

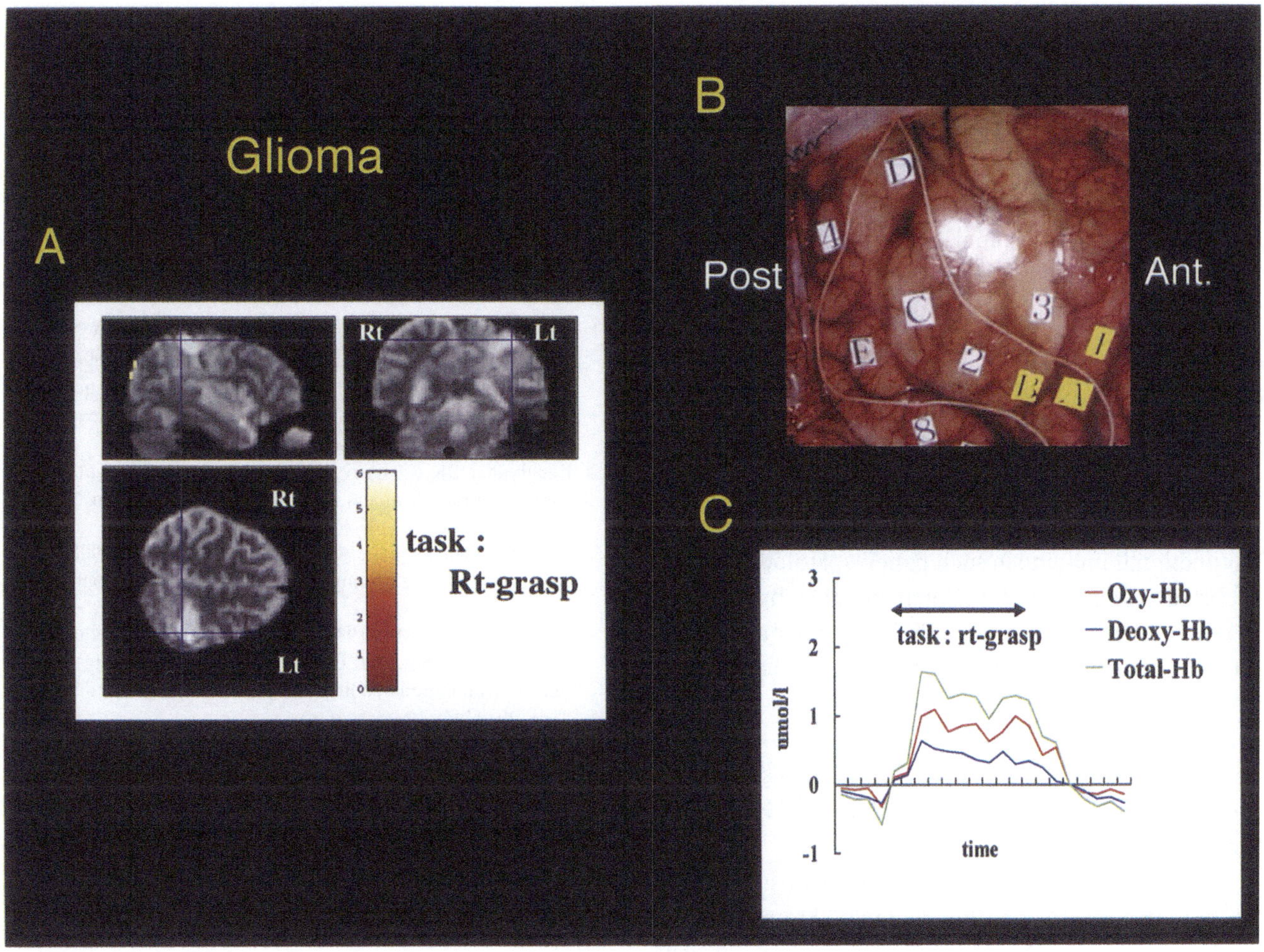

Fig. 2. (A) Threshold activated areas superimposed onto T_2 images of the glioma patient (right motor task). (B) Operative view of the cortical mapping of the glioma patient during awake surgery. (C) CBO changes in the PSMC of the glioma patient during contralateral motor tasks measured by NIRS. Note that Deoxy-Hb increased in the patient. The ordinates indicate concentration changes of Oxy-Hb, Deoxy-Hb, and Total-Hb. Horizontal thick bars indicate the period of the task

during the task in all of the cerebral ischemia, indicating the presence of rCBF increase in response to neuronal activation (Fig. 1B right). A focal concentration of Deoxy-Hb in the cerebral ischemia increased during the entire course of the task.

In two patients with glioma, BOLD-fMRI demonstrated very small activation areas in the PSMC on the lesion side during the task (Fig. 2A); activation in the PSMC was confirmed by cortical mapping during awake surgery in this patient (Fig. 2B). NIRS demonstrated an increase of Deoxy-Hb during the task, associated with increases of Oxy-Hb and Total-Hb; this CBO change was similar to that in the cerebral ischemia (Fig. 2C).

Discussion

In the present study, BOLD-fMRI demonstrated only limited area of activation in the PSMC on the lesion side in six cerebral ischemic patients and two glioma patients. In these patients, the focal concentration of Deoxy-Hb increased in the PSMC during the task, associated with increases of Oxy-Hb and Total-Hb, whereas Deoxy-Hb decreased in the normal matured brain. The increases in Oxy-Hb and Total-Hb suggest that the rCBF in the PSMC was increased during the motor task. The increase of Deoxy-Hb was evident during the entire course of the task. Thus, it differs from the Deoxy-Hb rise occurring within a few seconds after the start of neuronal activation [8], or the "post-stimulus overshoot" of Deoxy-Hb [9].

All patients could perform the task similarly to the normal subjects. In addition, the significant increases of Oxy-Hb and Total-Hb indicated an activation of the PSMC in the patients; in the glioma patients, activation in the PSMC was confirmed by cortical mapping or cortico-spinal motor evoked potential during surgery. These observations suggest that the BOLD-fMRI did not image the activation area correctly. Such a failure might be caused by the SPM, because it images the activation areas by detecting increases of BOLD signals (i.e. decreases of Deoxy-Hb) during activation.

Recently, several BOLD-fMRI studies on patients with brain tumors have cast doubt on the reliability of its functional imaging in such patients. Holodney *et al.* reported a failure of functional imaging by BOLD-fMRI in the patients with glioblastoma multiforme [10]. A number of BOLD-fMRI studies have demonstrated similar false negative activations in brain tumors. The present study suggests that the false negative activation of BOLD-fMRI is caused by an increase of focal concentration of Deoxy-Hb at the activation area.

The underlying physiological mechanism of the Deoxy-Hb increase in the damaged brain remains unclear. We hypothesize that the cerebral oxygen consumption of the cerebral ischemic patients may have been increased more by neuronal activity than in the normal adults. The oxygen consumption could increase in the activation area, if the activation area utilized the energy metabolism that requires oxygen, such as oxidation of lactate. In the ischemic patients studied, SPECT demonstrated a low baseline rCBF and a reduced cerebrovascular reactivity in the PSMC on the lesion side. These findings imply that metabolic acidosis occurred at rest, associated with the presence of lactate in the PSMC of the patients. Under such conditions, when oxygen was supplied through an increase of rCBF in response to neuronal activation, the lactate might have been oxidized, resulting in an increase of Deoxy-Hb. Indeed, in vitro studies have demonstrated that lactate can serve as an aerobic energy substrate for neuronal activation induced by glutamate [11].

In summary, the evoked-CBO changes occurring in the damaged brain differed from those in the normal brain. This indicates that BOLD-fMRI may overlook activation areas in the damaged brain. Finally, we have observed similar Deoxy-Hb rises in the PSMC during motor tasks in the patients with tremor (unpublished data). Such cortical activities may be overlooked by conventional BOLD-fMRI.

References

1. Atlas SW, Howard RS, Maldjian J, Alsop D, Detre JA, Listerud J, D'Esposito M, Judy KD, Zager E, Stecker M (1996) Functional magnetic resonance imaging of regional brain activity in patients with intracerebral gliomas: findings and implications for clinical management. Neurosurgery 38: 329–338
2. Buchbinder BR, Cosgrove GR (1998) Cortical activation MR studies in brain disorders. Magn Reson Imaging Clin N Am 6: 67–93
3. Fox PT, Raichle ME, Mintun MA, Dence C (1988) Nonoxidative glucose consumption during focal physiologic neural activity. Science 241: 462–464
4. Fox PT, Raichle ME (1986) Focal physiological uncoupling of cerebral blood flow and oxidative metabolism during somatosensory stimulation in human subjects. Proc Natl Acad Sci USA 83: 1140–1144
5. Ogawa S, Tank DW, Menon R, Ellerman JM, Kim S-G, Merkle H, Ugurbil K (1992) Intrinsic signal changes accompanying sensory stimulation: Functional brain mapping with magnetic resonance imaging. Proc Natl Acad Sci USA 89: 5951–5955
6. Sakatani K, Xie Y, Lichty W, Li S, Zuo H (1998) Language-activated cerebral blood oxygenation and hemodynamic changes of the left prefrontal cortex in poststroke aphasic patients: A near infrared spectroscopy study. Stroke 29: 1299–1304
7. Greenberg MS (1997) Handbook of neurosurgery, 4th edn. Greenberg Graphic Inc, USA, pp 52–53
8. Malonek D, Grinvald A (1996) Interactions between electrical activity and cortical microcirculation revealed by imaging spectroscopy: Implications for functional brain imaging. Science 272: 551–554
9. Heekeren HR, Obrig H, Wenzel R, Eberle K, Ruben J, Villringer K, Kurth R, Villringer A (1997) Cerebral hemoglobin oxygenation during sustained visual stimulation: a near-infrared spectroscopy study. Philos Trans R Soc Lond B Biol Sci 352(1354): 743–750
10. Holodny AI, Schulder M, Liu WC, Wolko J, Maldjian JA, Kalnin AJ (2000) The effect of brain tumors on BOLD functional MR imaging activation in the adjacent motor cortex: Implications for image-guided neurosurgery. Am J Neuroradiol 21: 1415–1422
11. Schurr A, Miller JJ, Payne RS, Rigor BM (1999) An increase in lactate output by brain tissue serves to meet the energy needs of glutamate-activated neurons. J Neurosci 19: 34–39

Correspondence: Kaoru Sakatani, M.D., D.M.Sc., Ph.D., Department of Neurosurgery, Nihon University School of Medicine, 30-1 Oyaguchi-Kamimachi, Itabashi-ku, Tokyo 173-8610, Japan. E-mail: sakatani@med.nihon-u.ac.jp

Acta Neurochir (2003) [Suppl] 87: 63–69

Acetazolamide vasoreactivity in persistent vegetative state and vascular dementia evaluated by transcranial harmonic perfusion imaging and Doppler sonography

T. Shiogai[1], **M. Koshimura**[1], **C. Uebo**[1], **M. Makino**[2], **T. Mizuno**[2], and **K. Nakajima**[2]

[1] Department of Clinical Neurosciences, Kyoto Takeda Hospital, Kyoto, Japan
[2] Department of Neurology, Kyoto Prefectural University of Medicine, Kyoto, Japan

Summary

To clarify the pathophysiological differences of the cerebrovascular reserve capacity in relation to cerebral cognitive impairments between persistent vegetative state (PVS) and vascular dementia (VD), we evaluated acetazolamide (ACZ) vasoreactivity testing by transcranial harmonic perfusion imaging (HPI) and Doppler sonography (TCD).

Methods. The subjects were 11 adult patients with severe cognitive impairments (4 PVS, 7 VD). TCD mean velocity (Vm) in the middle and posterior cerebral artery (MCA, PCA) and peak intensity (PI), area under curve (AUC), and mean transit time (MTT) analyzed by HPI time-intensity curves in the bilateral temporal lobe (TL), basal ganglia (BG), and thalamus (Th) were evaluated before and after ACZ administration. Resting values and relative changes (%Δ) of TCD and HPI parameters were compared between PVS and VD.

Results. a) resting values: There were no significant differences between the two groups. b) Vasoreactivity: 1) PVS: %Δ Vm decreased in the left PCA and MCA. %Δ PI/AUC/MTT decreased in the left TL and bilateral BG. 2) VD: %Δ PI/AUC decreased in the right TL. %Δ MTT tended to decrease in the right side.

Conclusion. ACZ vasoreactivity tests by transcranial HPI and TCD allowed bedside, non-invasive, quantitative evaluation of the pathophysiology of cognitive function impairment and treatments, in relation to cerebrovascular reserve capacity in PVS and VD.

Keywords: Acetazolamide vasoreactivity; persistent vegetative state; vascular dementia; transcranial harmonic perfusion imaging; Doppler sonography.

Introduction

"Wakefulness without awareness" is a typical feature of persistent vegetative state (PVS) [13] where there is a loss of awareness of the external world, cognitive function is severely disturbed or virtually lost. A definition and clinical features of PVS have been established [1] and medical aspects of PVS have been reviewed and defined in more detail [30]. However, misdiagnosis of PVS based on clinical findings still occurs [2, 4]. Furthermore, the differentiation between concepts of dementia and disturbance of consciousness in the chronic stage is not always easy and sometimes both conditions are confused and mixed up [14].

Diagnostic clinical criteria of vascular dementia (VD) for research studies have been proposed [23]. However, the sensitivity and specificity for detecting VD is not high [10], and probably due to the heterogeneity of vascular dementia syndromes and pathologic subtypes; including ischemic and hemorrhagic strokes, cerebral hypoxic ischemic events, and senile leukoencephalopathic lesions [23]. The importance of neuroimaging studies to support clinical findings has been emphasized. In particular, vasoreactivity tests induced by acetazolamide (ACZ) or CO_2 have been tried in order to find a distinction between VD and Alzheimer's dementia [15, 17, 19]. Vasoreactivity tests utilizing transcranial Doppler (TCD) and CBF measurements have been introduced for evaluating cerebrovascular reserve capacity in ischemic cerebrovascular diseases [12, 18, 20, 21]. However, correlation of vasoreactivity between TCD and CBF measurements is not always high and superiority of the modalities has not been established [6, 18, 20, 21]. Therefore, studies of both TCD and CBF measurements are necessary for evaluating vasoreactivity precisely. However, techniques for CBF measurements, such as single photon emission CT (SPECT), xenon-enhanced CT (Xe-CT), and positron emission CT (PET), are complex, labor-intensive, and impossible to use during treatments at bedside. Recently, transcranial harmonic imaging utilizing ultrasound contrast agents and color duplex sonography (CDS) has been

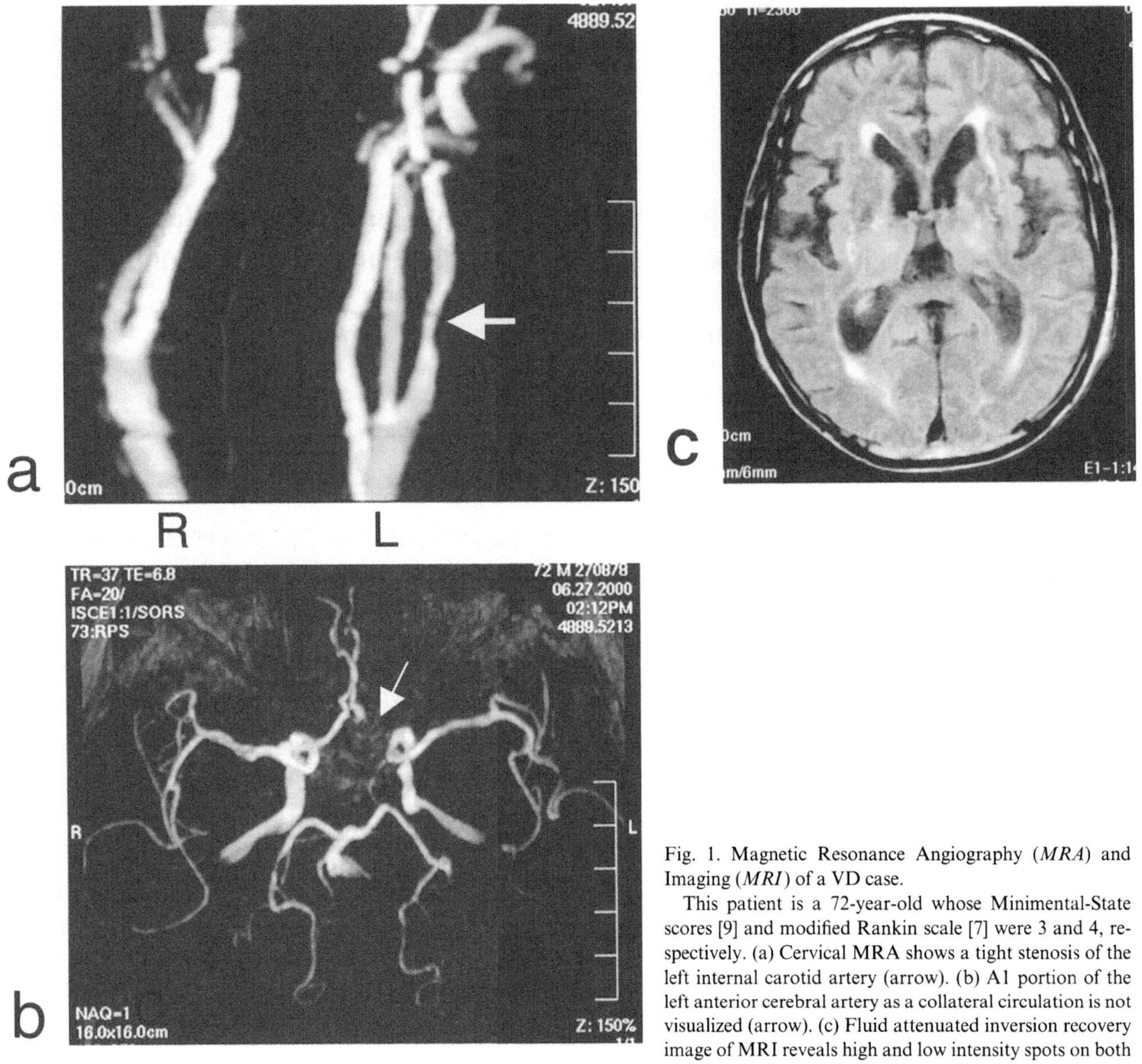

Fig. 1. Magnetic Resonance Angiography (*MRA*) and Imaging (*MRI*) of a VD case.

This patient is a 72-year-old whose Minimental-State scores [9] and modified Rankin scale [7] were 3 and 4, respectively. (a) Cervical MRA shows a tight stenosis of the left internal carotid artery (arrow). (b) A1 portion of the left anterior cerebral artery as a collateral circulation is not visualized (arrow). (c) Fluid attenuated inversion recovery image of MRI reveals high and low intensity spots on both basal ganglia and the periventricular high intensity area

introduced for repeatable bedside measurements of brain tissue perfusion [22, 24, 26, 27].

With respect to the CBF study for PVS patients, global or hemispheric CBF reduced [25, 30]. However, vasoreactivity of PVS in relation to the cognitive impairments has not been elucidated.

The aim of this study is to clarify the pathophysiological differences of the cerebrovascular reserve capacity in relation to cerebral cognitive impairments between PVS and VD, and to this end we evaluated ACZ vasoreactivity testing by transcranial harmonic perfusion imaging (HPI) and TCD.

Patients and methods

The subjects were 11 patients with severe cognitive impairments (4 PVS, 7 VD; age 41–85 years, mean 69). The age, primary diagnosis, major site of cerebral and/or vascular lesions based on imaging studies of CT, magnetic resonance imaging and angiography (Fig. 1), and CDS; scores of mini-mental state examination [9], modified Rankin grade [7], and state scale for chronic state of disturbed consciousness [16] of the two groups are shown in Table 1. All PVS patients had restricted cerebrovascular and anoxic causes. Time-averaged mean velocity (Vm) in the middle and posterior cerebral artery (MCA, PCA) was measured and HPI utilizing an echo-contrast agent (Levovist® 2.5 g) was evaluated on the axial plane involving the temporal lobe (TL), basal ganglia (BG), and thalamus (Th). Time-intensity curves of the HPI were analyzed for 3 regions of

Table 1. *Clinical characteristics of the patient groups**

	VD	PVS
n	7	4
Age		
– Mean ± SD (range)	73 ± 10 (57–85)	61 ± 27 (41–72)
Primary diagnosis		
– Cerebral infarction		
lacunae	4	
hemispheric	2	
angioma	1	
– Moyamoya disease		1
– Cerebral hemorrhage		1
– Subarachnoid hemorrhage		1
– Anoxia		1
Major site of cerebral and/or vascular lesions based on imaging studies**		
– R	5	1
– L	1	2
– Diffuse or none	1	1
Mini-Mental State Examination		
– Mean (range)	9 (3–19)	0 (0)
Modified Rankin grade		
– Mean (range)	4 (3–4)	5 (5)
State Scale Score for Chronic State of Disturbed Consciousness†		
– Mean (range)	10 (10)	6 (4–7)
Duration after a major cerebral insult		
– Mean (range)	16 (4–40)	13 (2–43)

* *VD* patients with vascular dementia, *PVS* persistent vegetative state patients, ** CT, MRI, MRA, and/or Color Duplex Sonography.

† State Scale for Chronic State of Disturbed Consciousness: 1. Physiological stability (normal cardiovascular and respiratory function). 2. Spontaneous eye opening. 3. Sleep-wake cycle. 4. Spontaneous movements of the oral cavity and pharynx (swallowing). 5. Spontaneous involuntary movements of the extremities. 6. Spontaneous changes in the expression muscles. 7. Spontaneous involuntary vocalization. 8. Interest in circumstances. 9. Voluntary purposeful movements. 10. Coherent verbalization (speech).

interest; the TL, BG, and Th, on both sides (Fig. 2). The TCD and HPI were evaluated before (resting state) and after (15 and 30 minutes on the right and left sides, respectively) the injection of ACZ (Diamox® 500 mg) (Fig. 2). Vasoreactivity was compared between the two groups in terms of resting values and relative changes (%Δ) of Vm and TIC parameters of peak intensity (PI), area under curve (AUC), and mean transit time (MTT).

Utilizing a Student t-test, statistical significance was set at p values less than 0.05.

Results

a) Resting values (Table 2): There were no significant differences between the two groups in Vm and HPI parameters. Vm in the left MCA of PVS tended to decrease, despite PI, AUC, and MTT tending to decrease in the right TL of PVS and left TL of VD.

b) Vasoreactivity (Table 2): 1) PVS: %Δ Vm decreased significantly in the left PCA ($p < 0.05$) and tended to decrease in the left MCA. There was also a tendency of decrease %Δ Vm in the right MCA. %Δ PI/AUC/MTT tended to decrease in the left TL and bilateral BG. %Δ PI/AUC also tended to decrease in the left Th. 2) VD: %Δ PI/AUC decreased in the right TL. %Δ MTT showed negative mean values in the right side.

Discussion

CBF and vasoreactivity evaluated by TCD and HPI

Expected close relationships of resting values and vasoreactivity based on vascular distribution [29] between measurements in the MCA by TCD and in the TL and BG by HPI, and between measurements in the PCA by TCD and in the Th by HPI, were not always observed. However, there were some correlated tendencies between TCD and HPI parameters. Particularly in the PVS group, tendencies of %Δ Vm decrease were obvious in the left MCA and slight in the right MCA, and coincided with decreased %Δ PI and %Δ AUC in the left TL and BG and right BG. Despite an observable significant decrease of %Δ Vm in the left PCA, there was a decreased tendency of %Δ PI in the left Th. However, in the VD group, despite %Δ PI and %Δ AUC showing negative values in the right TL, decreased %Δ Vm in the right MCA was not observed.

Correlation of vasoreactivity evaluated by TCD and CBF measurements based on SPECT or Xe-CT are not always high ($r = 0.39–0.67$) in normal subjects and stroke patients [18, 24, 28]. There was no correlation between flow velocity observed by TCD and CBF in normal subjects [6] and patients with carotid occlusive lesions [21]. These discrepancies of measured vasoreactivity between TCD and HPI are probably due to topographical differences between the main trunks of MCA and PCA and small regions of interest in the TL, BG, and Th in our series (Figs. 2c and 2d).

In other relevant points, there are methodological problems of quantitative measurements of brain tissue perfusion by HPI [27]; e.g. 1) bone- and depth-dependent attenuation of the ultrasound signals, 2) shadowing caused by different absorption and scattering [3], 3) bubble saturation due to a non-linear rela-

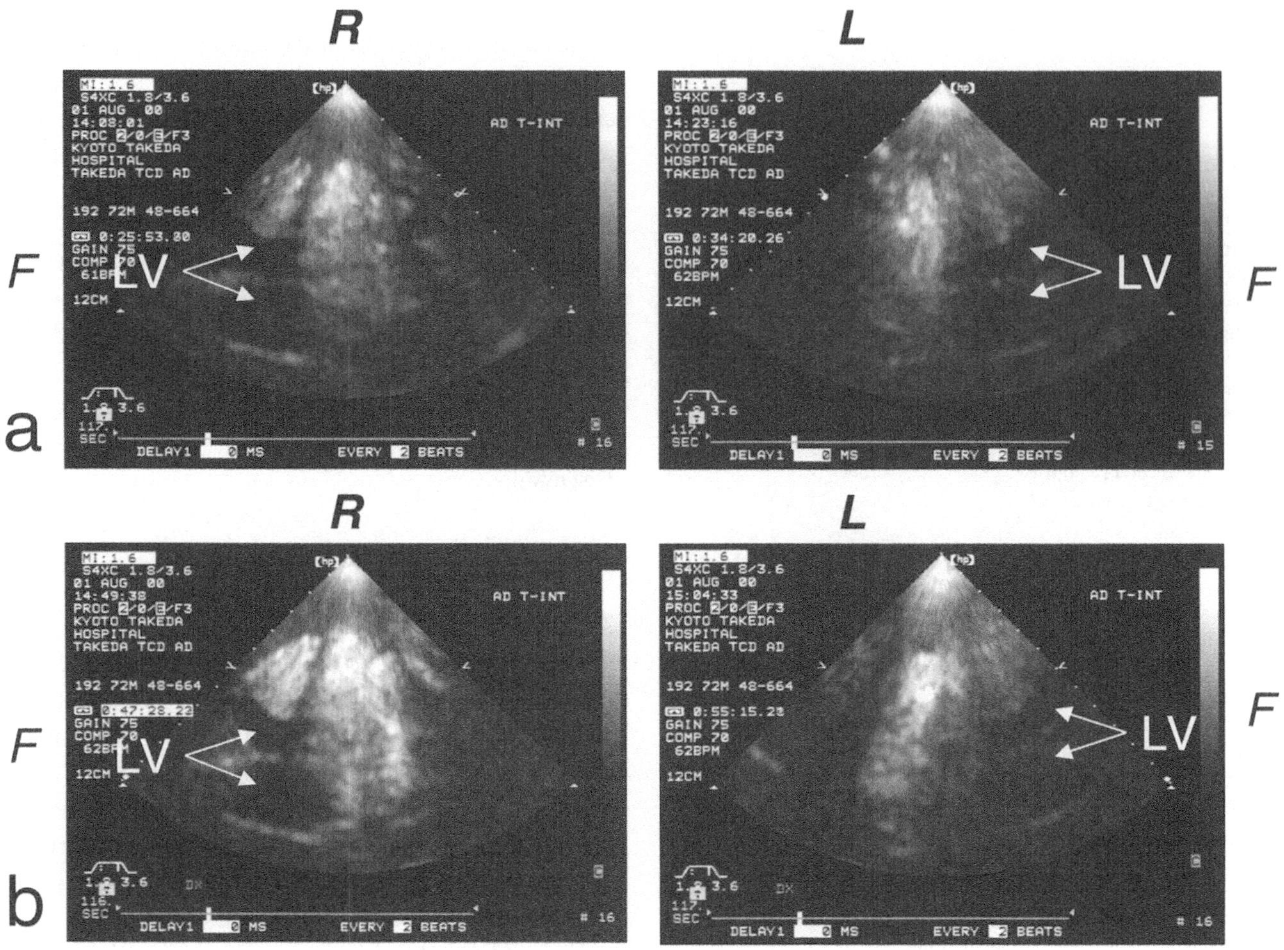

Fig. 2. HPI of the case in Fig. 1. LV indicates bilateral anterior horn of the lateral ventricle. F indicates the frontal side.

Post-contrast HPI via right (*R*) and left (*L*) temporal windows demonstrated before (a) and after (b) ACZ injection. Preserved vasoreactivity indicated by increased high intensity areas after ACZ injection (b) was obvious in the right side involving the TL, BG and Th in comparison with the left side, which involved mainly BG.

Time-intensity curve analyses in the three regions of interest (*ROI*) in the temporal lobe (*TL*), basal ganglia (*BG*), and the thalamus (*Th*) are performed via a right temporal window (*R*) before (c) and after (d) ACZ injection

tionship between microbubble concentration and image intensity [5], 4) fixation of ultrasonic probe, etc.

CBF and vasoreactivity in VD and PVS

Resting values measured by TCD and HPI show no significant differences between the PVS and VD groups. There were no consistent tendencies of Vm decrease in the left MCA and of PI, AUC, or MTT decrease in the right TL of PVS and left TL of VD. Inconsistent decreases of CBF are probably attributed to heterogeneous distribution of CBF caused by primary vascular and anoxic-ischemic brain damage.

Vasoreactivity of %Δ PI and %Δ AUC in the VD group was impaired in the right TL. Lateral differences of %Δ MTT were detected in the TL, BG, and Th. These are probably due to extracranial vascular stenotic lesions (Figs. 1 and 2) or cerebral lesions that were frequently detected in the right side of this series (Table 1). Vasoreactivity impairments in cases of extracranial stenotic lesions have been pointed out [12, 21]. On the other hand, impairments of vasoreactivity in VD are reported not only in cases of large hemispheric infarct but also lacunar infarct [8, 17, 19] or Binswanger's disease [15]. However, vasoreactivity induced by hypocapnea in so-called multi-infarct de-

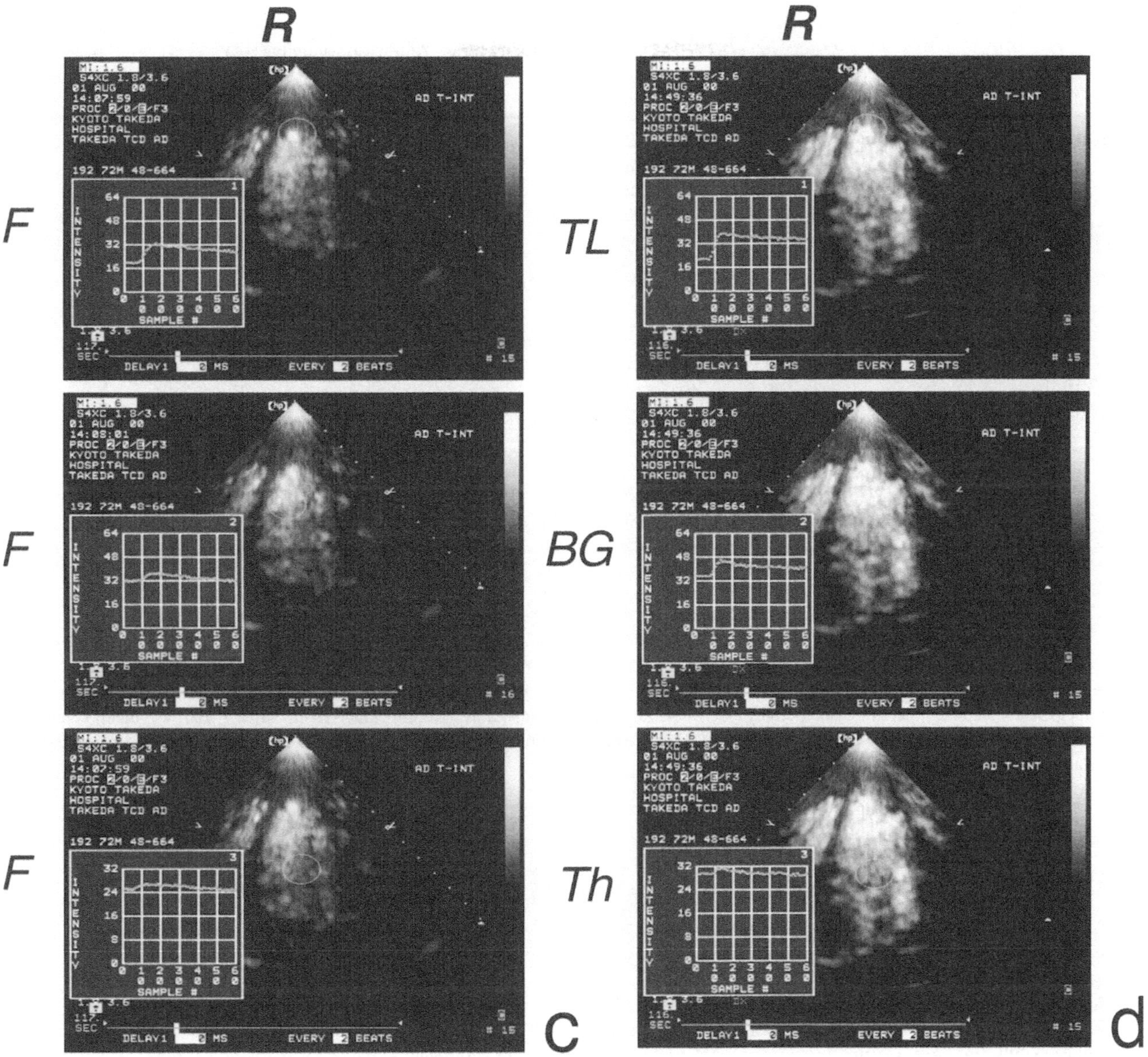

Fig. 2. (continued)

mentia is not impaired [11]. In our observation and analysis of published reports, it seems obviously difficult to clearly determine if vasoreactivity impairments are caused by primary cerebral and/or vascular lesions or the vascular dementia itself. Further studies are essential for this clarification.

Vasoreactivity impairments of PVS are more extensive than those of VD. These findings are probably related to severe cognitive impairments in PVS. Vasoreactivity in PVS cases has not been investigated. In our series, preserved vasoreactivity in the Th was identified. This result is presumably related to the fact that CBF, cerebral metabolic rate of oxygen (CMRO2) and glucose decreased in both the cerebral and cerebellar cortex, while there is a tendency toward preserved CMRO2 in the thalamus and brainstem in PVS [25].

Further studies are recommended for ACZ vasoreactivity tests for VD and PVS patients that are intended to clarify its relationship to therapeutic and prognostic contributions, such as during and after pharmacological, physical, and electrical [31] stimulation of residual neuronal function.

Table 2. *Resting values and acetazolamide (ACZ) vasoreactivity evaluated by TCD and HPI*

		Resting values		p values*	ACZ vasoreactivity (%Δ)		p values*
		VD (n = 7)	PVS (n = 4)		VD (n = 7)	PVS (n = 4)	
Vm (cm/s)	MCA (R)	50.8 ± 11.2	53.8 ± 12.9	ns	28.6 ± 22.2	13.0 ± 25.4	ns
	PCA (R)	30.1 ± 12.4	31.7 ± 13.6	ns	19.1 ± 20.2	21.3 ± 16.3	ns
	MCA (L)	58.1 ± 13.0	46.2 ± 5.6	ns	16.8 ± 18.9	1.8 ± 2.8	ns
	PCA (L)	37.6 ± 20.5	43 ± 25.1	ns	33.8 ± 22.4	−7.14 ± 17.0	$p < 0.05$
PI (dB)	TL (R)	11.0 ± 4.8	8.18 ± 2.4	ns	−15.4 ± 28.1	28.4 ± 42.2	ns ($p < 0.07$)
	BG (R)	7.64 ± 3.7	8.78 ± 3.8	ns	29.3 ± 111.7	12.5 ± 52.9	ns
	Th (R)	4.07 ± 2.3	5.88 ± 4.5	ns	26.1 ± 89.5	65.7 ± 120	ns
	TL (L)	7.74 ± 4.9	11.9 ± 7.1	ns	21.6 ± 51.0	−36.7 ± 20.9	ns ($p < 0.06$)
	BG (L)	7.31 ± 4.3	8.83 ± 4.3	ns	54.6 ± 93.7	−9.1 ± 20.1	ns
	Th (L)	3.68 ± 3.1	3.3 ± 1.6	ns	4.94 ± 36.8	3.39 ± 36.2	ns
AUC (dB × s)	TL (R)	804 ± 466	497 ± 243	ns	−9.45 ± 47.3	36.4 ± 58.4	ns
	BG (R)	512 ± 284	662 ± 386	ns	62.1 ± 243	−10.8 ± 52.9	ns
	Th (R)	289 ± 158	454 ± 537	ns	23.8 ± 127	40.2 ± 92.8	ns
	TL (L)	500 ± 418	895 ± 562	ns	99.3 ± 172	−58.5 ± 26.5	ns
	BG (L)	430 ± 360	519 ± 370	ns	152 ± 276	−7.15 ± 24.4	ns
	Th (L)	176 ± 217	145 ± 64.8	ns	105 ± 244	36.9 ± 101	ns
MTT (s)	TL (R)	59.9 ± 6.7	54.4 ± 4.0	ns	−2.79 ± 20.9	1.62 ± 16.9	ns
	BG (R)	57.1 ± 9.8	57.6 ± 8.0	ns	−0.8 ± 43.8	−11.3 ± 17.4	ns
	Th (R)	56.1 ± 8.5	44.3 ± 20.1	ns	−11.7 ± 22.0	9.0 ± 17.5	ns
	TL (L)	47.6 ± 17.4	64.1 ± 5.7	ns	55.1 ± 96.3	−20.5 ± 18.6	ns
	BG (L)	50.2 ± 22.1	55.6 ± 9.8	ns	67.5 ± 143	−1.59 ± 20.2	ns
	Th (L)	50.8 ± 16.9	42.3 ± 16.6	ns	43.4 ± 127	50.3 ± 96.1	ns

* Student t-test, *ns* not significant. Vasoreactivity (%Δ) = value after ACZ injection − resting value/resting value × 100. *Vm* mean velocity, *PI* peak intensity, *AUC* area under curve, *MTT* mean transit time, *MCA* middle cerebral artery, *PCA* posterior cerebral artery, *TL* temporal lobe, *BG* basal ganglia, *Th* thalamus, *R* right, *L* left.

References

1. ANA Committee on Ethical Affairs (1993) Persistent vegetative state: report of the American Neurological Association Committee on Ethical Affairs. Ann Neurol 33: 386–390
2. Andrews K, Murphy L, Munday R, Littlewood C (1996) Misdiagnosis of the vegetative state: retrospective study in a rehabilitation unit. BMJ 313: 13–16
3. Bos LJ, Piek JJ, Spaan JA (1996) Effects of shadowing on the time-intensity curves in contrast echocardiography: a phantom study. Ultrasound Med Biol 22: 217–227
4. Childs NL, Mercer WN, Childs HW (1993) Accuracy of diagnosis of persistent vegetative state. Neurology 43: 1465–1467
5. Claassen L, Seidel G, Algermissen C (2001) Quantification of flow rates using harmonic grey-scale imaging and an ultrasound contrast agent: an in vitro and in vivo study. Ultrasound Med Biol 27: 83–88
6. Dahl A, Russell D, Rootwelt K, Nyberg-Hansen R, Kerty E (1995) Cerebral vasoreactivity assessed with transcranial Doppler and regional cerebral blood flow measurements. Dose, serum concentration, and time course of the response to acetazolamide. Stroke 26: 2302–2306
7. De Haan R, Limburg M, Bossuyt P, van der Meulen J, Aaronson N (1995) The clinical meaning of Rankin 'handicap' grades after stroke. Stroke 26: 2027–2030
8. De Reuck J, Decoo D, Hasenbroekx MC, Lamont B, Santens P, Goethals P, Strijckmans K, Lemahieu I (1999) Acetazolamide vasoreactivity in vascular dementia: a positron emission tomographic study. Eur Neurol 41: 31–36
9. Folstein MF, Folstein SE, McHugh PR (1975) "Mini-mental state". A practical method for grading the cognitive state. J Psychiatr Res 12: 189–198
10. Gold G, Giannakopoulos P, Montes-Paixao Junior C, Herrmann FR, Mulligan R, Michel JP, Bouras C (1997) Sensitivity and specificity of newly proposed clinical criteria for possible vascular dementia. Neurology 49: 690–694
11. Hachinski VC, Iliff LD, Zilhka E, Du Boulay GH, McAllister VL, Marshall J, Russell RW, Symon L (1975) Cerebral blood flow in dementia. Arch Neurol 32: 632–637
12. Heiss WD, Podreka I (1993) Role of PET and SPECT in the assessment of ischemic cerebrovascular disease. Cerebrovasc Brain Metab Rev 5: 235–263
13. Jennett B, Plum F (1972) Persistent vegetative state after brain damage: A syndrome in search of a name. Lancet 1: 734–737
14. Kaneko M (1998) Clinical Observation of dementia particularly correlation between the disturbance of consciousness in the chronic stage. Proc Soc Treat Coma 7: 11–15
15. Kuwabara Y, Ichiya Y, Otsuka M, Masuda K, Ichimiya A, Fujishima M (1992) Cerebrovascular responsiveness to hypocapnea in Alzheimer's dementia and vascular dementia of the Binswanger type. Stroke 23: 594–598
16. Matsui T (1997) Proposal of new scoring system for chronic state of disturbed consciousness. Proc Soc Treat Coma 5: 53–59
17. Meyer JS, Shirai T, Akiyama H (1996) Neuroimaging for differentiating vascular from Alzheimer's dementias. Cerebrovasc Brain Metab Rev 8: 1–10
18. Muller M, Voges M, Piepgras U, Schimrigk K (1995) Assessment of cerebral vasomotor reactivity by transcranial Doppler ultrasound and breath-holding. A comparison with acetazolamide as vasodilatory stimulus. Stroke 26: 96–100

19. Pavics L, Grunwald F, Reichmann K, Horn R, Kitschenberg A, Hartmann A, Menzel C, Schomburg AG, Overbeck B, Csernay L, Biersack HJ (1999) Regional cerebral blood flow single-photon emission tomography with 99mTc-HMPAO and the acetazolamide test in the evaluation of vascular and Alzheimer's dementia. Eur J Nucl Med 26: 239–245
20. Piepgras A, Schmiedek P, Leinsinger G, Haberl RL, Kirsch CM, Einhaupl KM (1990) A simple test to assess cerebrovascular reserve capacity using transcranial Doppler sonography and acetazolamide. Stroke 21: 1306–1311
21. Pindzola RR, Balzer JR, Nemoto EM, Goldstein S, Yonas H (2001) Cerebrovascular reserve in patients with carotid occlusive disease. Stroke 32: 1811–1887
22. Postert T, Muhs A, Meves S, Federlein J, Przuntek H, Buttner T (1998) Transient response harmonic imaging: an ultrasound technique related to brain perfusion. Stroke 29: 1901–1907
23. Roman GC, Tatemichi TK, Erkinjuntti T, Cummings JL, Masdeu JC, Garcia JH, Amaducci L, Orgogozo JM, Brun A, Hofman A, Moody DM, O'Brien MD, Yamaguchi T, Grafman J, Drayer BP, Bennet DA, Fishe M, Ogata J, Kokmen E, Bermejo F, Wolf PA, Gorelink PB, Bick KL, Pajeau AK, Bell MA, Decarli C, Curebras A, Korezyn AD, Bogousslavsky J, Hartmann A, Sheinberg P (1993) Vascular dementia: diagnostic criteria for research studies. Report of the NINDS-AIREN International Workshop. Neurology 43: 250–260
24. Seidel G, Algermissen C, Christoph A, Claassen L, Vidal-Langwasser M, Katzer T (2000) Harmonic imaging of the human brain. Visualization of brain perfusion with ultrasound. Stroke 31: 151–154
25. Shiogai T (1995) Cerebral blood flow and metabolism and cerebrospinal fluid neurotransmitters in the persistent vegetative state. Soc Treat Coma 3: 37–46
26. Shiogai T, Tsunezuka C, Koshimura M, Takayasu N, Nagakane Y, Nakase T, Makino M, Nakajima K, Furuhata H (2000) Comparative visualization of brain tissue perfusion in various neurological disorders utilizing contrast-enhanced harmonic imaging in integrated backscatter and power Doppler methods. Cerebrovasc Dis 10 [Suppl] 1: 44 [abstract]
27. Shiogai T, Tsunezuka C, Ohara T, Imai K, Makino M, Nakajima K, Furuhata (2001) Clinical significance of transcranial contrast-enhanced harmonic perfusion imaging (HPI) as a quantitative method of measuring cerebral blood flow. Neurotraumatology 24: 77–82 [in Japanese]
28. Sorteberg W, Lindegaard KF, Rootwelt K, Dahl A, Nyberg-Hansen R, Russell D, Nornes H (1989) Effect of acetazolamide on cerebral artery blood velocity and regional cerebral blood flow in normal subjects. Acta Neurochir (Wien) 97 (3–4): 139–145
29. Tatu L, Moulin T, Bogousslavsky J, Duvernoy H (1998) Arterial territories of the human brain: cerebral hemispheres. Arterial territories of the human brain: cerebral hemispheres. Neurology 50: 1699–1708
30. The Multi-Society Task Force on PVS (1994) Medical aspects of the persistent vegetative state (First of two parts). N Engl J Med 330: 1499–1508
31. Tsubokawa T, Yamamoto T, Katayama Y, Hirayama T, Maejima S, Moriya T (1990) Deep-brain stimulation in a persistent vegetative state: follow-up results and criteria for selection of candidates. Brain Injury 4: 315–327

Correspondence: Toshiyuki Shiogai, M.D., Ph.D., Department of Clinical Neurosciences, Kyoto Takeda Hospital, Minami-kinutacho 11, Nishi-nanajo, Shimogyo-ku, Kyoto 600-8884, Japan. E-mail: shiogait@pop11.odn.ne.jp

Acta Neurochir (2003) [Suppl] 87: 71–74

Impairment of motor function after frontal lobe resection with preservation of the primary motor cortex

C. Fukaya, Y. Katayama, K. Kobayashi, M. Kasai, H. Oshima, and **T. Yamamoto**

Department of Neurological Surgery and Division of Applied System Neuroscience, Nihon University School of Medicine, Tokyo, Japan

Summary

We investigated the clinical course and characteristics of the motor deficits in patients who underwent surgical resection of the frontal lobe for tumorous lesions. Only patients who met the following criteria were included in the present study: 1) postoperative MRI revealed that resection of the frontal lobe involved the area closely adjacent to the primary motor cortex, but 2) the D wave of the corticospinal MEP did not decrease in amplitude below 50% of the original level during surgery. The extent of resection was classified into 4 groups. In Group A (6 cases), resection was limited within the area above the superior frontal sulcus and posterior to a line vertical to the line connecting the anterior and posterior commissures at the anterior commissure (AC vertical line). Resection was extended anterior to the AC vertical line in Group B (4 cases) or below the superior frontal sulcus in Group C (5 cases). In Group D (3 cases), resection was extended to both of these two boundaries. Severe motor paresis and/or apraxia of the upper and lower extremities were noted in all patients of Group D immediately after surgery. A complete recovery in the lower extremity was observed in these patients, while disturbance in the fine movements of the upper extremity remained for more than 1 year after the surgery. Disturbance in the fine movements and/or apraxia of the upper extremity were observed immediately after surgery in 2 of the Group A patients (33%), 2 of the Group B patients (50%) and 3 of the Group C patients (60%). However, a rapid recovery occurred in these patients, and only a subtle or mild disturbance remained for more than 1 year after the surgery in one of the Group B and one of the Group C patients. Permanent and severe motor deficit is rarely induced when resection of the frontal lobe is limited to only the SMA proper (corresponding roughly to Group A), the SMA proper and pre-SMA (corresponding roughly to Group B), or the SMA proper and premotor cortex (corresponding roughly to Group C), insofar as the primary motor cortex is preserved. Disturbance in fine movements of the upper extremity is frequently induced for the long term when wide areas of the SMA proper, pre-SMA as well as premotor cortex are resected altogether (corresponding roughly to Group D).

Keywords: Fine movement; functional recovery; premotor cortex; primary motor cortex; supplementary motor cortex.

Introduction

A large area of the frontal lobe located anterior to the primary motor cortex is involved in the execution and control of motor function [1, 3–5]. In non-human primates, this area is defined as being located anterior to the primary motor cortex, limited by the cingulate sulcus, and extending onto the superior convexity of the cerebral hemisphere. Deficits in motor function are therefore induced either transiently or permanently by surgical resection of this area in humans. Little detail has yet been given, however, regarding the clinical course and characteristics of the motor deficits in relation to the precise extent of surgical resection of this area. Much controversy remains concerning the localization of motor function in the frontal lobe of humans.

It is impossible to define anatomically the primary motor cortex during surgery. It is possible, however, to detect function of the primary motor cortex during surgery based on the corticospinal motor evoked potential (MEP). The corticospinal MEP represents a direct impulse (D wave) mediated by fast corticospinal tract neurons. Since the fast corticospinal tract neurons originate largely in the primary motor cortex, preservation of the primary motor cortex is assured by an intact D wave. We investigated the clinical course and characteristics of the motor deficits in patients who underwent surgical resection of the frontal lobe for tumorous lesions and whose D wave was largely preserved during the surgery.

Clinical materials and methods

A total of 18 patients were included in this prospective study. They comprised 10 male and 8 female patients with ages ranging from 23 to 63 years (mean age: 40 years). All patients harbored gliomas involving the dominant side (8 cases) or nondominant side (10 cases). Each patient was assessed neurologically over a 1-year period following surgery. No obvious motor deficits were noted in these patients preoperatively.

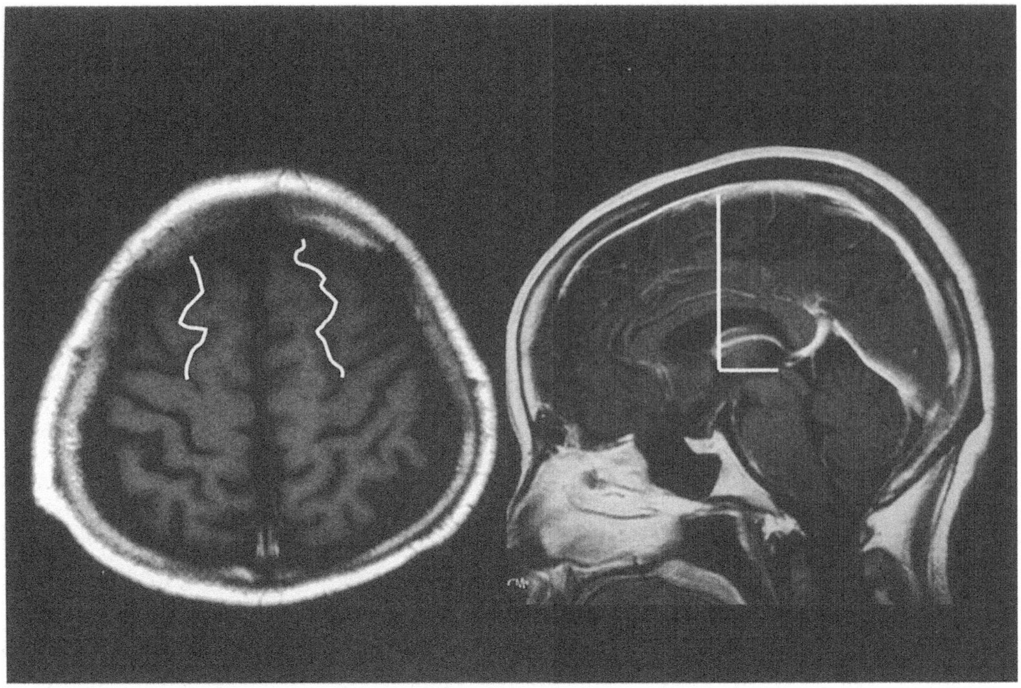

Fig. 1. Two boundaries for classifying the extent of resection. One was the superior frontal sulcus (right), and the other was a line vertical to the line connecting the anterior and posterior commissures at the anterior commissure (AC vertical line; left)

During surgery, the corticospinal MEP, recorded from the spinal epidural space in response to direct cortical stimulation of the primary motor cortex, was monitored in all patients. The technique of corticospinal MEP monitoring was as described elsewhere [2]. All patients underwent pre- and postoperative magnetic resonance (MR) imaging studies. The postoperative MR images were obtained within 72 hours. The MR imaging included enhancement with gadolinium-diethylenetriamine pentaacetic acid. Only patients who met the following criteria were included in the present study: 1) postoperative MRI revealed that resection of the frontal lobe involved the area closely adjacent to the primary motor cortex, but 2) the D wave of the corticospinal MEP did not decrease in amplitude below 50% of the original level during surgery.

Based on the extent of resection, the patients were classified into 4 groups according to 2 anatomically defined boundaries: one was the superior frontal sulcus and the other was a line vertical to the line connecting the anterior and posterior commissures at the anterior commissure (AC vertical line). The AC vertical line is supposed to divide the supplementary motor area (SMA) proper and the pre-SMA (Fig. 1). In Group A (6 cases), resection was limited within the area above the superior frontal sulcus and posterior to the AC vertical line. Resection was extended anterior to the AC vertical line in Group B (4 cases) or below the superior frontal sulcus in Group C (5 cases). In Group D (3 cases), resection was extended to both of these two boundaries.

Results

In Group A, 2 patients (33%) presented with disturbance in fine movements of the upper extremity immediately after surgery (Table 1). They recovered completely from their motor deficit within 1 year. In Group B, 2 patients (50%) presented with disturbance in fine movements of the upper extremity immediately after surgery (Table 1). One of them recovered completely from his motor deficit within 1 year. In the other patient, a subtle disturbance in fine movements of the upper extremity remained for more than 1 year. In Group C, 3 patients (60%) presented with disturbance in fine movements and apraxia of the upper extremity (Table 1). In 2 of them, rapid recovery was noted and the motor deficits disappeared within 3 months. Mild disturbance of fine movements remained

Table 1. *Relationship between extent of resection and disturbance in movements of the upper extremity*

	Total n	Disturbance in movements of the upper extremity	
		Immediately after surgery	One year after surgery
Group A	6	2 33%	0
Group B	4	2 50%	1 25%
Group C	5	3 60%	1 20%
Group D	3	3 100%	3 100%

for more than 1 year in the other patient, but his activity of daily living was not so disturbed. In Group D, all patients (100%) presented with severe motor paresis and/or apraxia of the upper and lower extremities immediately after surgery (Table 1). They could not move their upper and lower extremities at all during the immediate period after the surgery. A rapid recovery in the lower extremity and gradual recovery in the upper extremity were observed in these patients. Their motor deficit in the lower extremity improved greatly so that all patients became able to walk within 1 month. However, disturbance in the fine movements of the upper extremity remained for more than 1 year after the surgery. It is worth mentioning that none of the patients investigated in the present study suffered permanent gait disturbance.

Discussion

The present results demonstrate that, even if the primary motor cortex is preserved, varying degrees of motor deficits are induced immediately after surgery by resection which involves the SMA proper, the pre-SMA or the premotor cortex. The present findings also indicate, however, that permanent and severe motor deficit rarely remains for the long term when resection is limited to the SMA proper (corresponding roughly to Group A), or to the pre-SMA (corresponding roughly to Group B) or the premotor cortex (corresponding roughly to Group C) in addition to the SMA proper, insofar as the primary motor cortex is preserved. This appears to be consistent with previous clinical observations as well as experimental studies on animals. Zentner *et al.* [5] reported that only a subtle motor deficit remained after resection of the SMA in humans. Similarly, Halsband and Passingham [1] found that severe motor deficit did not remain after resection of the premotor cortex in non-human primates.

The roles played by the SMA and the premotor cortex in complex motor tasks have been assessed previously in animal experiments and from clinical observations. It has been shown that, depending on the extent of SMA resection, motor tasks are impaired in association with weakness and/or apraxia of the extremities. Such deficits are believed to be related to the function of the SMA that deals with the planning and initiation of motor activity [5]. Mushiake *et al.* [4] noted that the SMA is critically involved in the preparation of internally remembered motor sequences but not in sensory-guided sequential activities. Thus, the SMA may play a key role in volitional processes guided by memory.

The concept of premotor syndrome was established by Fulton. According to his report, mastered voluntary movements such as buttoning up a shirt are disturbed without motor weakness following injury of the premotor cortex. Such symptoms are often closely akin to the apraxia presenting after injury of the parietal cortex. This finding points to the existence of a strong connection between the premotor cortex and the parietal cortex playing a role in the process of sensory input. Experiments employing non-human primates have indicated that the role of the premotor cortex is to integrate the information of the visual and other sensory inputs necessary for voluntary movement [4]. In other words, the premotor area works mainly for sensory-guided movements.

The present results also indicate that disturbance in fine movements of the upper extremity is frequently induced for the long term when wide areas of the SMA proper, the pre-SMA and the premotor cortex are resected altogether (Group D). Awake surgery under local anesthesia cannot provide information regarding the postoperative functional recovery for the long term. When a surgeon decides on the extent of resection, the central, precentral and superior frontal sulci are employed as anatomical landmarks to locate the primary and premotor cortices. In contrast, the AC vertical line is impossible to identify during surgery. In our experience, the distance from the central sulcus to the AC vertical line is approximately 3~4 cm. This information appears to be important for the surgeon in deciding the extent of resection.

Conclusion

Permanent and severe motor deficit is rarely induced when resection of the frontal lobe is limited to the SMA proper and pre-SMA, or the SMA proper and premotor cortex, insofar as the primary motor cortex is preserved. Disturbance in fine movements of the upper extremity is frequently induced for the long term when wide areas of the SMA proper, pre-SMA as well as premotor cortex are resected altogether.

References

1. Halsband U, Passingham RE (1985) Premotor cortex and the conditions for movement in monkeys (Macaca fascicularis). Behav Brain Res 18: 269–277

2. Katayama Y, Tsubokawa T, Maejima S, Hirayama T, Yamamoto T (1988) Corticospinal direct response in humans: identification of the motor cortex during intracranial surgery under general anesthesia. J Neurol Neurosurg Psychiatry 51: 50–59
3. Lim SH, Dinner DS, Pilly PK, Lüders H, Morris HH, Klem G, Wyllie E, Award IA (1994) Functional anatomy of the human supplementary sensorimotor area: results of extraoperative electrical stimulation. J Clin Neurophysiol 91: 179–193
4. Mushiake H, Inase M, Tanji J (1991) Neuronal activity in the primate premotor, supplementary, and precentral motor cortex during visually guided and internally determined sequential movements. J Neurophysiol 66: 705–718
5. Zentner J, Hufnagel A, Pechstein U, Wolf HK, Schramm J (1996) Functional results after receptive procedures involving the supplementary motor area. J Neurosurg 85: 542–549

Correspondence: Chikashi Fukaya, M.D., Department of Neurological Surgery and Division of Applied System Neuroscience, Nihon University School of Medicine, Itabashi-ku, Tokyo 173-8610, Japan. E-mail: @med.nihon-u.ac.jp

Acta Neurochir (2003) [Suppl] 87: 75–78

Rehabilitation technique facilitates association cortices in hemiparetic patients: functional MRI study

M. Maruishi[1], **H. Muranaka**[1], **M. Miyatani**[2], **J. Kawahara**[2], **T. Tsuji**[3], **S. Noda**[3], and **T. Kajima**[4]

[1] Hiroshima Prefectural Rehabilitation Center, Hiroshima, Japan
[2] Graduate School of Education, Hiroshima University, Hiroshima, Japan
[3] Graduate School of Engineering, Hiroshima University, Hiroshima, Japan
[4] Department of Radiology, Hiroshima Prefectural University of Health and Welfare, Hiroshima, Japan

Summary

We used fMRI to study brain activation with facilitative rehabilitation techniques (passive hand movements and visual feedback) in two patients with subcortical lesions. Two tasks were given in a sequence. The first task (*trial 1*) was repetitive hand grasping by the paretic hand at a rate of 0.5 Hz with the eyes closed. The second task (*trial 2*), the facilitative rehabilitation technique, included *task 1* plus support by a trainer to move the paretic hand with the eyes open to get visual feedback of the movement. The data were analyzed by a subtractive method. When *task 1* was subtracted from *task 2*, it was found that the bilateral visual cortex, contralateral premotor cortex and posterior parietal cortex were involved with the passive hand movement and visual feedback. These facilitative rehabilitation techniques may integrate networks between sensory information and motor commands, and lead to functional reorganization.

Keywords: Functional MRI; rehabilitation; association cortex.

Introduction

Hemiparesis remains a major deficit in patients with stroke. Recovery after stroke can be prolonged well past the period of the acute stage. This suggests that mechanisms other than simple reperfusion or resolution of the injury response might contribute to recovery. Facilitative rehabilitation techniques have been believed to improve the function of paretic limbs after stroke. However, it is not clear how these techniques facilitate functional recovery in the cerebral cortex. In this study, we report preliminary data obtained by functional magnetic resonance imaging (fMRI). Two patients with subcortical lesions and 4 healthy volunteers underwent an fMRI to assess the effect of facilitative rehabilitation techniques.

Subjects and methods

We used an fMRI to study brain activation with facilitative rehabilitation techniques (passive hand movements and visual feedback) in two patients with subcortical lesions. Both patients demonstrated a subcortical infarction at the right corona radiata, sparing the cerebral cortex.

Patient 1, a 57-year-old man, experienced a small infarction in the right corona radiata. Six weeks after the ictus (at the time of the fMRI), he was capable of generating very weak muscle contraction of the left hand flexors. Patient 2, a 60-year-old woman, had a small infarction in the right corona radiata. Twelve weeks after the ictus (at the time of the fMRI), she was also capable of generating very weak flexor muscle contractions of the left hand. Figure 1 shows the extent of the lesions in these patients.

We studied 4 healthy control volunteers (mean age 24 years, range 22 to 26 years, 3 women and 1 man) who had not previously suffered from recognized neurological disease. Ethical approval was obtained from the Ethics Committee of Hiroshima Prefectural Rehabilitation Center, and informed consent was obtained prior all studies.

Subjects were required to place their hands on their abdomens. Two tasks were given in a sequence. The first trial (*trial 1*) was repetitive hand grasping by the paretic hand at a rate of 0.5 Hz, with the eyes closed. The second trial (*trial 2*), the facilitative rehabilitation technique, was *trial 1* plus support by a trainer to move the paretic hands, with the eyes opened to get visual feedback of the movement.

The paradigms were organized into an ABCA pattern, in which A involved rest, B involved performing *trial 1*, and C involved performing *trial 2*, each lasting 60 seconds. The paradigms were performed three times with each subject.

Images were acquired with a Siemens/Symphony 1.5 T MRI scanner. BOLD images were obtained continuously by using the following parameters: repetition time 60 seconds, echo time 60 ms, 3-mm slice thickness, 40 slices, field of view 194 × 194, and 64 × 64 matrix.

Image processing and statistical analysis were carried out by using the SPM99 software. Motion correction was performed by using the realignment procedure, and spatial smoothing was applied before performing the statistical analysis. The images then were linearly

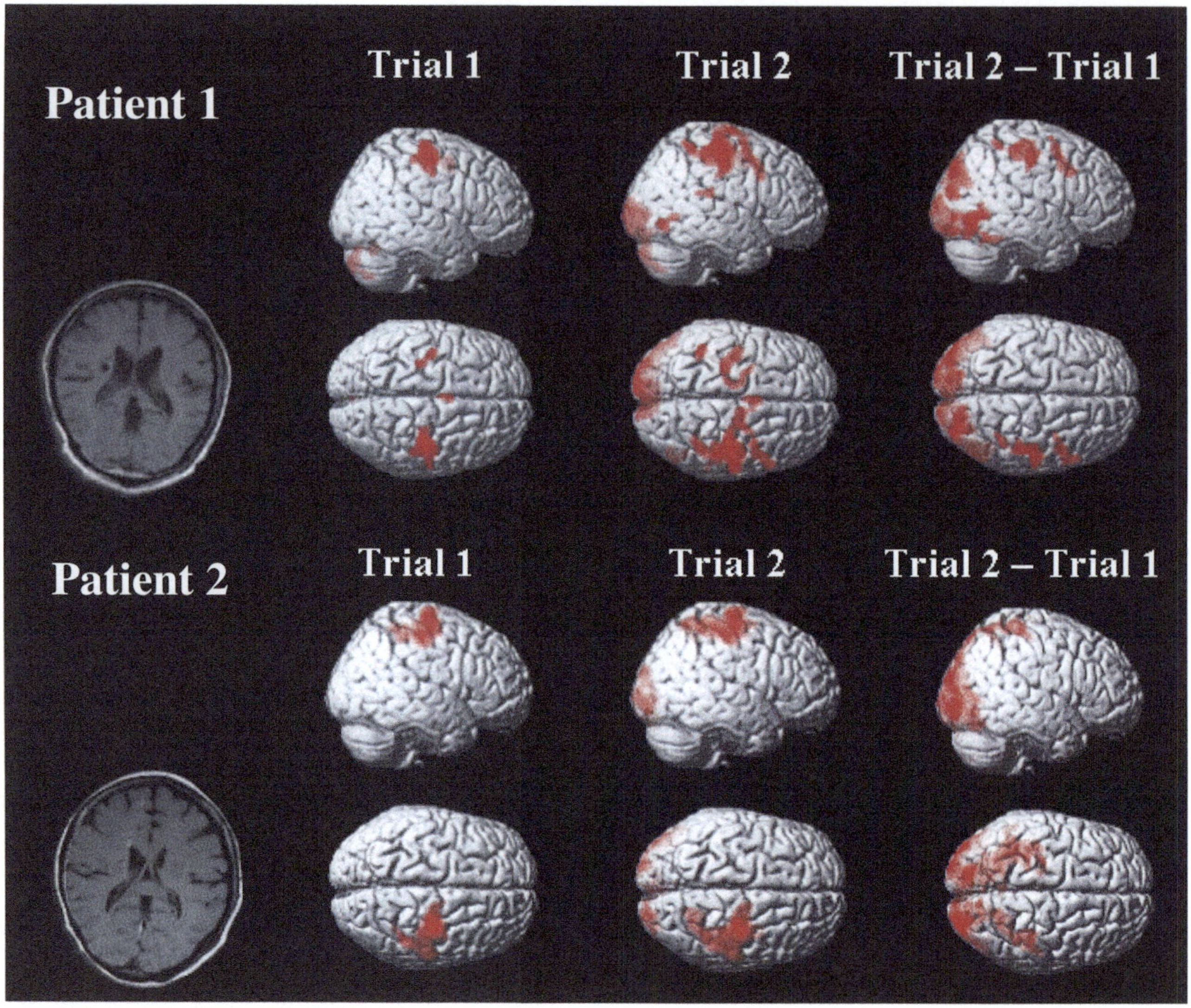

Fig. 1. Lesion location and functional MRI data for *Patient 1* and *Patient 2*. The first column illustrates high-resolution T1 weighted image. The second column illustrates activation associated with *trial 1*. The third column illustrates activation associated with *trial 2*. The fourth column illustrates activation associated with subtraction method between trials. Significant activation of posterior parietal cortex was shown in the subtraction method

transformed into a common stereotactic space by using a normalization procedure. The geometric center of the activation cluster was determined and expressed in Talairach coordinates in the average brain space. Activation maps were calculated by using a Student's parametric unpaired *t* test.

Results

Task-related cortical activations are shown for Patient 1 in Figure 1. In *trial 1*, activation of the bilateral primary sensorimotor cortex (SM1) and supplementary motor area (SMA) was detected. In *trial 2*, in addition to strong activation of the bilateral SM1, bilateral primary visual cortices (V1), visual association cortices (V2), contralateral posterior parietal cortex (PPC), and contralateral premotor cortex (PM) were also activated. The data were analyzed by a subtractive method. When *trial 1* was subtracted from *trial 2*, it was noted that the bilateral V1 and V2, and the contralateral PM and PPC were involved in the effect of passive hand movement and visual feedback.

For Patient 2, *trial 1* demonstrated a small activation of the contralateral SM1. In *trial 2*, in addition to strong activation of the contralateral SM1, bilateral V1 and V2 and the contralateral PPC were also acti-

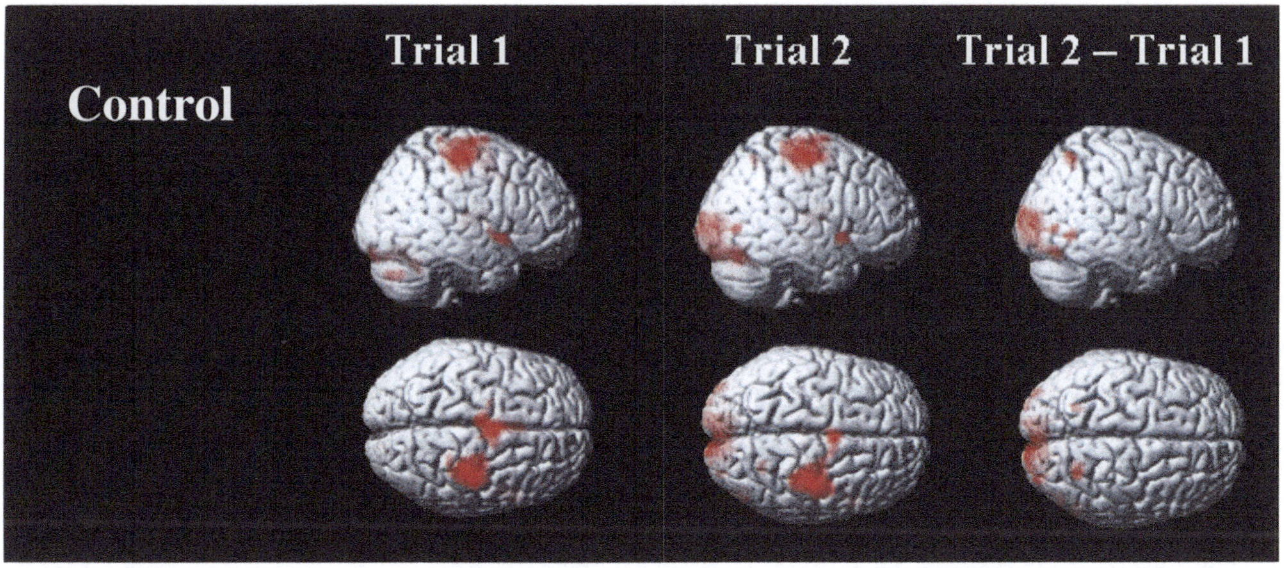

Fig. 2. Functional MRI data for 4 healthy control subjects. The first column illustrates activation associated with *trial 1*. The second column illustrates activation associated with *trial 2*. The third column illustrates activation associated with subtraction method between trials. Less activation of posterior parietal cortex was shown in the subtraction method

vated. Bilateral V1 and V2 and the contralateral PPC were found to be active when *trial 1* was subtracted from *trial 2*.

Figure 2 illustrates data from healthy control subjects. In *trial 1*, activation of the contralateral SM1 and SMA were detected. In *trial 2*, activation of the contralateral SM1 was similar to that of *trial 1*, *and* contralateral PPC were less activated. The contralateral PPC were found to be less active when the *trial 1* was subtracted from the *trial 2*.

Discussion

We used an fMRI to study brain activation with facilitative rehabilitation techniques (passive hand movements and visual feedback) in two patients with subcortical lesions. A subtraction method revealed that these techniques activated bilateral V1, V2, PPC, and the contralateral SM1 and PM. This pattern was not found in control subjects.

Injury-induced cortical reorganization is a widely recognized phenomenon. In contrast, there is less information on treatment-induced plastic changes in the human brain. The forced-use therapy was believed to be specifically effective in further improving recovery in stroke patients. Liepert *et al.* [2, 3] used focal transcranial magnetic stimulation to determine motor cortex plasticity during forced-use therapy in stroke patients. They suggested that the forced-use therapy enhanced motor cortex excitability and improved motor performance, but the question remains of what effect such an approach can have in the cortical network of motor learning.

The primary motor cortex has two levels of functional organization [1]. First, a low-level control system, the corticomotoneuronal cells, controls groups of muscles that can be brought together into task-specific combinations. Second, a higher-level control system, encodes more global features of the movement. Nelles *et al.* [5] used serial positron emission tomography to study training-induced (passive movement) brain plasticity after a severe hemiparetic stroke. After arm training, activation was not only found bilaterally in the PPC and PM, but also in the contralateral SM1. Their research suggested that arm training induced functional brain reorganization in bilateral sensory and motor systems. Our study also suggested that training-induced reorganization occurs in such higher-level control systems.

The execution of voluntary movement relies on sensorimotor transformation in which representations of the external environment are integrated into motor programs [1]. This integration is the product of premotor and primary motor areas operating in conjunction with sensory and association areas [4, 7]. An example of this has been seen in the communication between the parietal and motor areas during visually guided reaching and grasping. During reaching, neurons in the PPC construct a code for the direction of the movement but discharge later than the dorsal PM neurons to which they are connected [6]. During grasping, different neurons in the ventral PM of monkeys fire in connection with different hand actions and object shapes. The cells in the ventral PM thus seem to direct motor actions that can be guided by visual information about object shape received from the posterior parietal cortex [6].

In general, voluntary movements improve in speed and accuracy with repeated practice. This may reflect an optimization process in which the minimal circuits needed to accomplish a behavior are, with training, selected from redundant sensorimotor connections. Such a process could be responsible for the organization of a network between one group of cells and another, or between one area of the cortex and another [1].

Our research suggested that a similar process might have occurred with the facilitative rehabilitation techniques, in which sensorimotor transformation affected reorganization of the motor control network, enhanced motor cortex excitability, and improved motor performance.

Conclusions

Our study confirmed the effectiveness of a facilitative rehabilitation technique. These techniques may integrate networks between sensory information and motor commands, and lead to functional reorganization.

Acknowledgments

The authors wish to thank Michele Eisemann Shimizu, Hiroshima Prefectural University of Health and Welfare, for her assistance in preparing the manuscript.

References

1. Krakauer J, Ghez C (2000) Voluntary movement. In: Kandel ER, Schwartz JH, Jessell TM (eds) Principles of neural science IV. McGraw Hill, New York, pp 756–779
2. Liepert J, Miltner WH, Bauder H, Sommer M, Dettmers C, Taub E, Weiller C (1998) Motor cortex plasticity during constraint-induced movement therapy in stroke patients. Neurosci Lett 250: 5–8
3. Liepert J, Uhde I, Graf S, Leidner O, Weiller C (2001) Motor cortex plasticity during forced-use therapy in stroke patients: a preliminary study. J Neurol 248: 315–321
4. Mountcastle VB, Lynch JC, Georgopoulous A, Sakata H, Acuna C (1975) Posterior parietal association cortex of the monkey: command functions for operations within extrapersonal space. J Neurophysiol 38: 871–908
5. Nelles G, Jentzen W, Jueptner M, Muller S, Diener HC (2001) Arm training induced brain plasticity in stroke studied with serial positron emission tomography. Neuroimage 13: 1146–1154
6. Rizzolatti G, Luppino G, Matelli M (1998) The organization of the cortical motor system: new concepts. Electroencephalogr Clin Neurophysiol 106: 283–296
7. Rushworth MF, Johansen-Berg H, Young SA (1998) Parietal cortex and spatial-postural transformation during arm movements. J Neurophysiol 79: 478–482

Correspondence: Masaharu Maruishi, M.D., Hiroshima Prefectural Rehabilitation Center, 295-3, Taguchi, Saijo-cho, Higashihiroshima, 739-0036, Japan. E-mail: maruishi@rehab-hirohima.gr.jp

Neurosurgical intervention for functional recovery from neurological deficit: Part 1

Acta Neurochir (2003) [Suppl] 87: 81–83

Early use of intrathecal baclofen in brain injury in pediatric patients

M. S. Turner

Pediatric Neurosurgery, Indianapolis Neurosurgical Group, Inc., IN, USA

Summary

Background. The Food and Drug Administration (FDA) approved the use of intrathecal Baclofen for spasticity from traumatic brain injury in June of 1996 based on a Phase III clinical trial that documented efficacy in patients one year post injury. The FDA approval is only for patients who are one year post injury based on the Phase III study. We have found use of ITB in the first few months after injury very effective in a subgroup of severe traumatic brain injury (TBI) pediatric patients with spasticity, dystonia and autonomic storming following brain injury.

Method. The author's database of over 250 patients receiving intrathecal baclofen was reviewed. Retrospective chart review was undertaken of the 6 patients identified with brain injuries over a three year period that were given ITB less than one year post injury. The patients' diagnosis included asphyxia, traumatic brain injury and stroke. The ages were one year to fourteen years of age. ITB was offered after all conventional therapy had been exhausted and the patient's spasticity remained intractable. These patients also all experienced autonomic dysfunction with severe autonomic storms.

Findings. All of the patients had a decrease in their spasticity of at least 2 points on the Ashworth score. The autonomic storms ceased in all 6 patients when adequate dosages of ITB were reached. The patients could be weaned from all oral and intravenous medications for tone, storming and fever. Many of the patients became much more alert and interactive when the medications were stopped. Five of the six are still using their pump for their tone one to five years post implant, the sixth had the pump stopped at the parents' request for religious reasons.

Interpretation. Early use of ITB can play a significant role in the rehabilitation of brain injury in children.

Keywords: Dysautonomia; brain injury; baclofen; intrathecal.

Introduction

The Food and Drug Administration (FDA) labeling for use of intrathecal baclofen (ITB) in brain injury approves its use in patients one year post injury. The recommendation is based on data from the Phase III study sponsored by Medtronic to obtain approval for the new indication. The study was designed to minimize any question of spontaneous improvement in tone and therefore limited entry to patients that were at least one year post injury. The FDA approval is therefore limited to those patients on which data of efficacy and safety exist. There is no published data that ITB is more or less effective in patients less than one year post injury. Several authors have reported "good outcomes" with ITB used in patients that were less than 12 months post injury [1, 2].

Severe brain injury in pediatric patients can cause intractable spasticity, dystonia and a syndrome of marked sweating, fever, tachycardia and tachypnea referred to as autonomic storming or dysautonomia. These symptoms occurred in each of our patients. These dysautonomic symptoms did not respond to large doses of medications directed at the spasticity and autonomic storms. The storming and sedation from the medications used in an attempt to control the symptoms severely interfered with attempts at rehabilitation and also prevented discharge from the hospital. The symptoms frequently precipitated returns to the intensive care unit for possible sepsis, with marked tachycardia, tachypnea and fever, or for oversedation. ITB was instituted in these patients to control their spasticity and allow basic hygiene, seating and range of motion. We found that as the tone improved the patients, autonomic symptoms resolved as well. Similar findings have been reported by others [3, 5].

Methods

The charts of all patients that had received a programmable pump for ITB by the author were reviewed. Six patients that began ITB therapy less than one year after their injury were identified. The patients' initial medication list, Ashworth score, and vital signs (temperature, pulse, blood pressure) before and after institution of ITB therapy were tabulated. Outcome was measured by reduction in medication, improvement in tone measured by the Ashworth score, and reduction in episodes of storming, expressed as maximum heart rate, respiratory rate and temperature each day before and after the

Table 1.

Patient number	Diagnosis	Pre ITB Ashworth score[1]	Days post injury ITB trial started	ITB dose at response micrograms/day
1	DAI[2]	4.1	48	144
2	DAI SDH[3]	4.0	55	144
3	anoxia	4.2	38	192
4	anoxia	4.4	42	240
5	infarct	4.5	45	144
6	anoxia	4.5	33	288
Average		4.3	43	192

Notes: 1. Average Ashworth of all limbs tested. 2. DAI Diffuse Axonal Injury. 3. SDH Subdural hematoma.

therapy, and the baclofen dose required to achieve those improvements. All of the patients underwent a trial of baclofen using an implanted catheter and continuous infusion using an external pump before implantation of the programmable pump. All patients had a successful trial and underwent implantation of a Medtronic SynchroMed® programmable pump system.

Results

The diagnosis and initial Ashworth scores are outlined in Table 1. The patients with anoxia all demonstrated severe opisthotonic posturing with their occiputs nearly touching their heels. The posturing episodes could not be prevented by large doses of multiple medications, and only occasionally could be broken with large doses of narcotics.

Case report

CA was a 10 year old female that had undergone partial resection of a right thalamic pilocytic astrocytoma. She had a mild residual left hemiparesis. Six months later she presented with an acute obstruction of her left middle cerebral artery, with a dense right hemiparesis and obtundation. Ten days post admission she developed episodes of tachycardia, tachypnea, and fever with profuse sweating. Medications were added sequentially in an attempt to block the symptoms without success. Her medication list at the time of initiation of her ITB trial is in Table 2. ITB was initiated through a catheter and the initial dose was 150 micrograms/day. Figures 1–3 demonstrate the response to the ITB on her pulse, Ashworth score and respiratory rate. She had a Medtronic SynchroMed pump placed and was weaned off her other antispasmodic, narcotic and sedative medications. She became much more alert and began to actively participate in her therapies. She is now over five years post implantation and still benefits from ITB.

Table 2.

Medications received daily	
Baclofen	80 mg/day
Dantrium	200 mg/day
Clonidine	0.6 mg/day
Parlodel	30 mg/day
Ativan	12–24 mg/day
Chloral hydrate	1500 mg/day

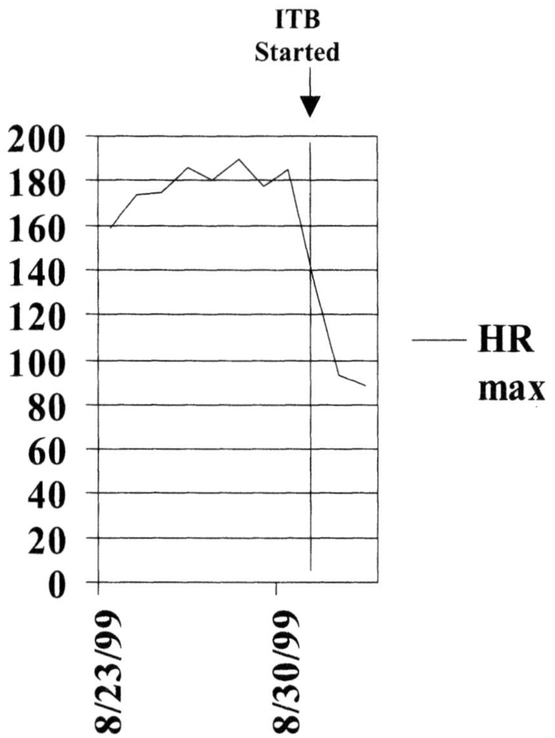

Fig. 1. Maximum heart rate each day

All of the patients had improvement in their tone, and abolishing of their autonomic storming. The dose of ITB to control storming and tone is listed in Table 1. The days post injury that ITB was instituted is in Table 1. All of the patients still require ITB therapy to maintain tone reduction. One pump was stopped at the parents' request for religious purposes. The patient still has severe spasticity. The earliest institution of ITB was at 33 days post injury.

Discussion

A computerized search of all literature concerning use of ITB in brain injury does not identify any report

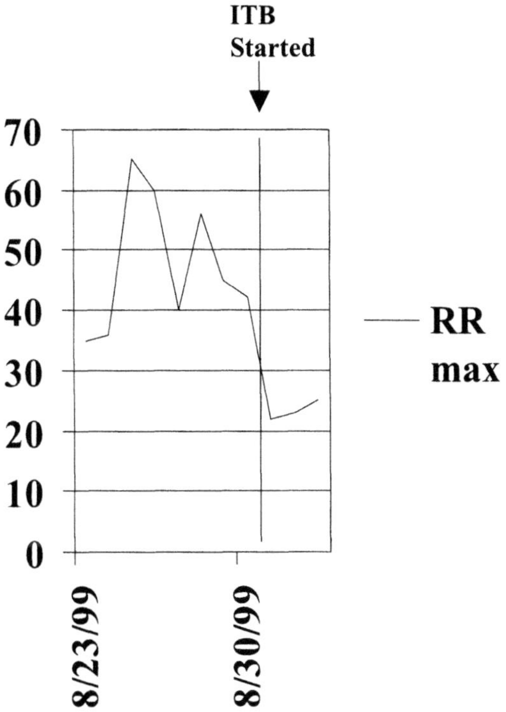

Fig. 2. Maximum respiratory rate each day

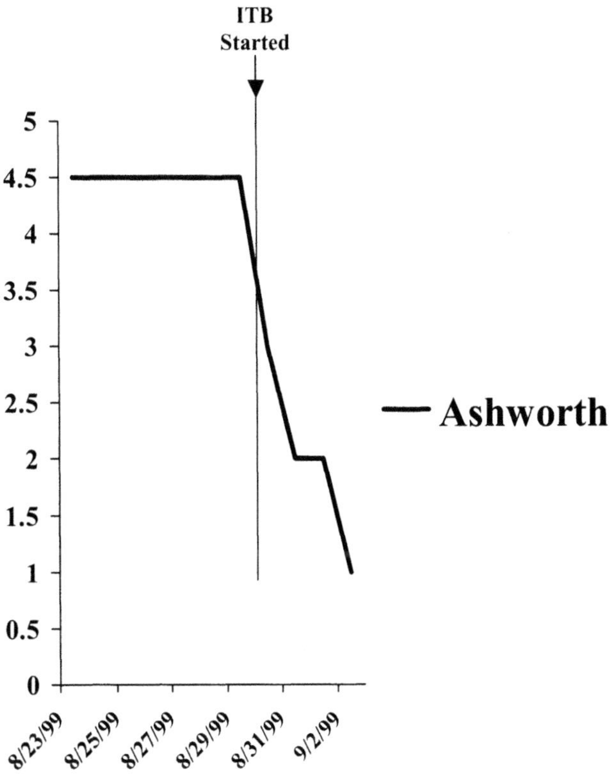

Fig. 3. Ashworth score each day

suggesting increased safety or efficacy when ITB is used one year post injury. Baclofen is a GABA agonist and there is a report in animals suggesting that GABA may inhibit neurotropic growth. This theoretically may interfere with brain repair after a brain injury [4].

We have identified a subgroup of patients with a variety of brain injuries that developed severe increases in tone with spasticity, dystonia and opisthotonus and the syndrome of autonomic storming in which early use of ITB was very effective in improving their tone and abolishing the dysautonomia. Other reports document similar findings [3, 5]. Several of the patients' spontaneous neurological improvements had been masked by the spasticity, dystonia and sedation from medications. Once ITB controlled the hypertonicity and the medications were weaned, the patients began purposeful interactions and were able to participate in their rehabilitation. None of our patients have had enough improvement in their tone to require or request removal of their pump. Further research may identify less severely impaired patients that may be candidates for use of ITB prior to one year.

References

1. Armstrong RW, Steinbok P, Cochrane DD, Kube SD, Fife SE, Farrell K (1997) Intrathecally administered baclofen for treatment of children with spasticity of cerebral origin. J Neurosurg 87(3): 409–414
2. Becker R, Alberti O, Bauer BL (1997, Mar) Continuous intrathecal baclofen infusion in severe spasticity after traumatic or hypoxic brain injury. J Neuro 244(3): 160–166
3. Becker R, Benes L, Sure U, Hellwig D, Bertalanaffy H (2000) Intrathecal baclofen alleviates autonomic dysfunction in severe brain injury. J Clin Neurosc 7(4): 316–319
4. Brailowski S, Knight RT, Blood K (1986) G-Aminobutyric acid-induced potentiation of cortical hemiplegia. Brain Res 362: 322–330
5. Cuny E, Richer E, Castel JP (2001) Dysautonomia syndrome in the acute recovery phase after traumatic brain injury: relief with intrathecal baclofen therapy. Brain Inj 15(10): 917–925

Correspondence: Michael S. Turner, M.D., Pediatric Neurosurgery, Indianapolis Neurosurgical Group, Inc., Indianapolis, IN, U.S.A. E-mail: TurnerLah@aol.com

Acta Neurochir (2003) [Suppl] 87: 85–87

Subtonsillar placement of auditory brainstem implant

Y. Seki[1], **N. Samejima**[1], **K. Kumakawa**[2], and **A. Komatsuzaki**[3]

[1] Department of Neurosurgery, Tokyo Kyosai Hospital, Tokyo, Japan
[2] Department of Otorhinolaryngology, Toranomon Hospital, Tokyo, Japan
[3] Department of Otorhinolaryngology, Chiba San-no Hospital, Chiba, Japan

Summary

A 26-year-old female with NF2 underwent removal of an acoustic schwannoma via extended retrosigmoid approach with subtonsillar placement of the auditory brainstem implant. The patient had already shown palsy of the lower cranial nerves on the opposite side due to previous surgery. Differing from conventional routes such as the translabyrinthine and the retrosigmoid, the subtonsillar approach [12] enabled us to observe the entire cochlear nucleus and avoid injury to the 9th and 10th cranial nerves when applying the electrode. Hearing in pure tone average was maintained almost at the preoperative level. We present a new use of this approach in placing the electrode and discuss its advantages.

Keywords: Auditory brainstem implant; neurofibromatosis type 2; subtonsillar approach.

Introduction

Patients who were deaf due to bilateral cochlear nerve dysfunction, such as neurofibromatosis type 2 (NF2) [3], could not have hearing restored until the introduction of the auditory brainstem implant (ABI) [4]. However, compared to the cochlear implant, with the ABI it is not so easy to properly place the electrode in the lateral recess because we can hardly observe the entire cochlear nucleus via the conventional translabyrinthine and retrosigmoid approaches. In this report, we performed placement of the electrode by a new route, the subtonsillar approach [12], which enables us to view the floor of the 4th ventricle including the whole cochlear nucleus.

Case presentation

The patient was 26 years old at the time of placement of the ABI. In August 1998, when she was 23 years old, she felt left hearing loss and 9 months later she showed an unsteady gait. In another hospital, MRI detected bilateral acoustic tumors, and she underwent 2-staged excision of the huge tumor on the left side.

Then she came to our hospital in February 2000. She showed paralyses of the left 5th, 7th, 8th, 9th and 10th cranial nerves, and severe left cerebellar signs. So we started rehabilitation mainly for cerebellar symptoms and also began observing the right side tumor, which was a few mm in extrameatal size. Hearing was 45 dB in pure tone average (PTA) and 20% in speech discrimination score (SDS). One and a half years later, although hearing remained almost at the same level (45 dB in PTA and 15% in SDS), the tumor grew to 15 mm. On November 14, 2001, we removed the tumor via a wide retrosigmoid craniectomy, including the opening of the foramen magnum and C1 laminectomy, followed by subtonsillar placement of the ABI. Along with conventional intraoperative monitoring of the electrically evoked ABR [11], subtonsillar placement allowed us to observe the cochlear nucleus and to directly measure the position of the electrode from the median sulcus (Fig. 1). After programming and stimulation of the ABI, 9 out of total 21 channels were available. Hearing was 48 dB in PTA, which was about the same as before the operation. She distinguishes various kinds of environmental sounds such as doorbells, phones ringing, dogs barking, and so on.

Discussion

At present, the ABI must be a godsend for patients deafened by bilateral cochlear nerve dysfunction. However, the performance of the ABI is still not

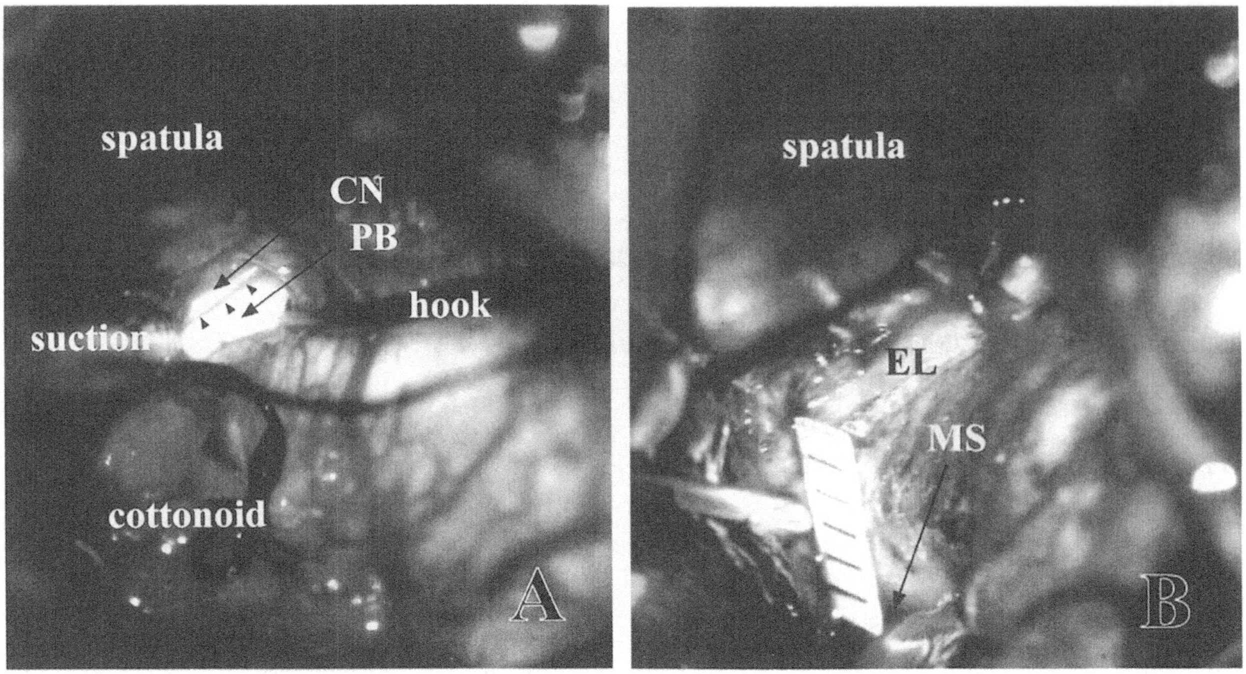

Fig. 1. Intraoperative photographs showing placement of the electrode by the subtonsillar approach with the patient being in the left-side down lateral position. (A) Spatula raises the right tonsil and the lateral recess is exposed by suction tip and hook. The vena nuclei cochlearis [5] (arrowheads) clearly separates the cochlear nucleus (*CN*) and pontobulbar body (*PB*). Cottonoid is covering the floor of the 4th ventricle. (B) Measure (6 mm) indicates the distance from the median sulcus (*MS*) to the medial edge of the electrode (*EL*)

very good compared to the cochlear implant. With a cochlear implant, many patients can use the telephone, but very few patients with the ABI can do so. One of the reasons for this is the difficulty in proper placement of the ABI electrode on the cochlear nucleus, as indicated in the recent report that 13 (14.8%) out of 88 ABI cases did not receive any auditory sensation [1]. Translabyrinthine and retrosigmoid approaches have so far been exclusively used for placement of the electrode [1, 10]. It is very hard to observe the entire cochlear nucleus in either approach. If one tries to do so, strong retraction of the flocculus and/or the choroid plexus, which may cause injury to the lower cranial nerves, or the introduction of the endoscope [2], is necessary. In our case, we had to preserve the facial nerve, as well as to avoid bilateral dysfunction of the 9th and 10th cranial nerves leading to tracheotomy. Therefore, we extended the conventional retrosigmoid craniectomy to a wide exposure, including the opening of the foramen magnum and C1 laminectomy, so as to enable subtonsillar placement of the electrode. In the subtonsillar approach (another name is cerebellomedullary fissure approach [6]), we can expose the whole cochlear nucleus by raising the tonsil and cutting the taenia from the foramen Magendie to the ipsilateral foramen Luschka without stretching the lower cranial nerves. Measurement between the conspicuous anatomical landmarks, such as the median sulcus to the auditory tubercle, is quite easy.

It is thought that, in the near future, a penetrating electrode array may be a promising means to improve the performance of the ABI because tonotopy of the cochlear nucleus is rather sharper from the surface towards the inside than on the surface [7, 8]. The subtonsillar approach may also be suitable for placing the depth electrodes. Other advantages of this approach are as follows: 1) the subtonsillar approach is not affected by adhesion caused by previous surgery for removal of the tumor, thus making a 2nd operation for the ABI feasible. 2) if necessary, this approach allows bilateral placement of the electrodes by exposing the whole floor of the 4th ventricle through raising both tonsils after bilateral cutting of the taenia [6].

Although it was originally developed mainly for removal of the dorsal brainstem tumor [6, 12], the subtonsillar approach can be beneficial for placement of the ABI, especially in conjunction with the meticulous measurement of the distances between the cochlear nucleus and surrounding structures on the floor of the 4th ventricle studied by Quester R and Schroeder R [9].

References

1. Ebinger K, Otto S, Arcaroli J, Staller S, Arndt P (2000) Multichannel auditory brainstem implant: US clinical trial results. J Larygol Otol 114 [Suppl] 27: 50–53
2. Friedland DR, Wackym PA (1999) Evaluation of surgical approaches to endoscopic auditory brainstem implantation. Laryngoscope 109: 175–180
3. Gutmann DH, Aylsworth A, Carey JC, Korf B, Marks J, Pyeritz RE, Rubenstein A, Viskochil D (1997) The diagnostic evaluation and multidisciplinary management of neurofibromatosis 1 and neurofibromatosis 2. JAMA 278: 51–57
4. Hitselberger WE, House WF, Edgerton BJ, Whitaker S (1984) Cochlear nucleus implant. Otolaryngol Head Neck Surg 92: 52–54
5. Klose AK, Sollmann WP (2000) Anatomical variations of landmarks for implantation at the cochlear nucleus. J Larygol Otol 114 [Suppl] 27: 8–10
6. Matsushima T, Inoue T, Inamura T, Natori Y, Ikezaki K, Fukui M (2001) Transcerebellomedullary fissure approach with special reference to methods of dissecting the fissure. J Neurosurg 94: 257–264
7. McCreery DG, Shannon RV, Moore JK, Chatterjee M (1998) Accessing the tonotopic organization of the ventral cochlear nucleus by intranuclear microstimulation. IEEE Trans Rehab Eng 6: 391–399
8. Otto SR, Brackmann DE, Hitselberger WE, Shannon RV (2001) Brainstem electronic implants for bilateral anacusis following surgical removal of cerebellopontine angle lesions. Otolaryngol Clin North Am 34: 485–499
9. Quester R, Schoeder R (1999) Topographic anatomy of the cochlear nuclear region at the floor of the fourth ventricle in humans. J Neurosurg 91: 446–476
10. Sollmann WP, Laszig R, Marangos N (2000) Surgical experiences in 58 cases using the Nucleus 22 multichannel auditory brainstem implant. J Larygol Otol 114 [Suppl] 27: 23–26
11. Waring MD (1995) Auditory brainstem responses evoked by electrical stimulation of the cochlear nucleus in human subjects. Electroenceph Clin Neurophysiol 96: 338–347
12. Ziyal IM, Sekhar LN, Sales E (1999) Subtonsillar-transcerebellomedullary approach to lesions involving the fourth ventricle, the cerebellomedullary fissure and the lateral brainstem. Br J Neurosurg 13: 276–284

Correspondence: Yojiro Seki, M.D., Department of Neurosurgery, Tokyo Kyosai Hospital, 2-3-8 Nakameguro, Meguro-ku, Tokyo, 153-8934, Japan. E-mail: yoseki@tkh.meguro.tokyo.jp

Acta Neurochir (2003) [Suppl] 87: 89–92

Diaphragm pacing with the spinal cord stimulator

T. Mitsuyama[1], T. Taira[1], N. Takeda[2], and T. Hori[1]

[1] Department of Neurosurgery, Neurological Institute, Tokyo Women's Medical University, Tokyo, Japan
[2] Department of Neurosurgery, Tokyo Metropolitan Okubo Hospital, Tokyo, Japan

Summary

Background. Diaphragm pacing with electrical stimulation of the phrenic nerve is an established treatment for central hypoventilation syndrome. The device, however, is not readily available, at least, in Japan. We applied the spinal cord stimulator for pain control to phrenic nerve stimulation. The purpose of this study is to evaluate the efficacy and feasibility of phrenic pacing with the compromise method.

Method and patients. We implanted a spinal cord stimulator in five patients with chronic central hypoventilation. The stimulation electrode was placed along the phrenic nerve in the neck, and the device was implanted in the anterior chest. We used the cyclic mode, and set the parameters at 1 second ramp up, 2 seconds on, 3 seconds off. The pulse width and the frequency were set at 150 µsec and 21 Hz respectively. The amplitude was adjusted to obtain sufficient tidal volume and to maintain $PaCO_2$ at around 40 mmHg.

Findings. During the follow-up period from 3 to 34 months (mean 23), stable and sufficient ventilation were observed in all patients without complications.

Interpretation. Although longer follow-up is necessary, diaphragm pacing with the spinal cord stimulator is feasible for treatment of central hypoventilation syndrome.

Keywords: Diaphragm pacing; spinal cord stimulator.

Introduction

Chronic hypoventilation due to dysfunction of the brainstem or the high cervical spinal cord poses a serious medicosocial problem. Patients with such hypoventilation are usually managed with the use of artificial ventilators. It has been shown for long time that diaphragm pacing with an implanted electric device to stimulate the phrenic nerve is a reasonable solution for such patients [1–5]. Almost all the patients with diaphragm pacing so far have been using a device specifically made for this purpose by Avery Laboratories Inc. (Commack, New York). Because this device is not readily available in our country, we applied a stimulator for spinal cord stimulation or deep brain stimulation for pain control to electrical stimulation of the phrenic nerves to pace the diaphragm.

Methods and patients

Patients

From March 2000 to June 2002, we performed diaphragm pacing using a stimulator for spinal cord stimulation in five patients with chronic hypoventilation and on an artificial ventilator due to brainstem or upper cervical cord dysfunction. There were one man and four women ranging in age from 35 to 63 years (mean age, 53.8 years). Table 1 summarizes the profile of these patients with regard to patient age, gender, the cause condition and duration of hypoventilation.

Determination of stimulation parameters

Before applying the stimulator in a clinical setting, the waveform of the output from the stimulation device was checked. According

Table 1. *Clinical characteristics of the patients*

Case	Age	Sex	Cause	Condition	Duration (months)
1	58	female	SAH (vertebral artery aneurysm)	sleep apnea (respirator at night)	6
2	58	female	operative complication (jugular foramen neurinoma)	totally dependent on respirator	4
3	63	female	brainstem hemorrhage (no operation)	totally dependent on respirator	8
4	55	male	brainstem hemorrhage (no operation)	totally dependent on respirator	3
5	35	female	operative complication (atlantaxial dislocation)	totally dependent on respirator	7

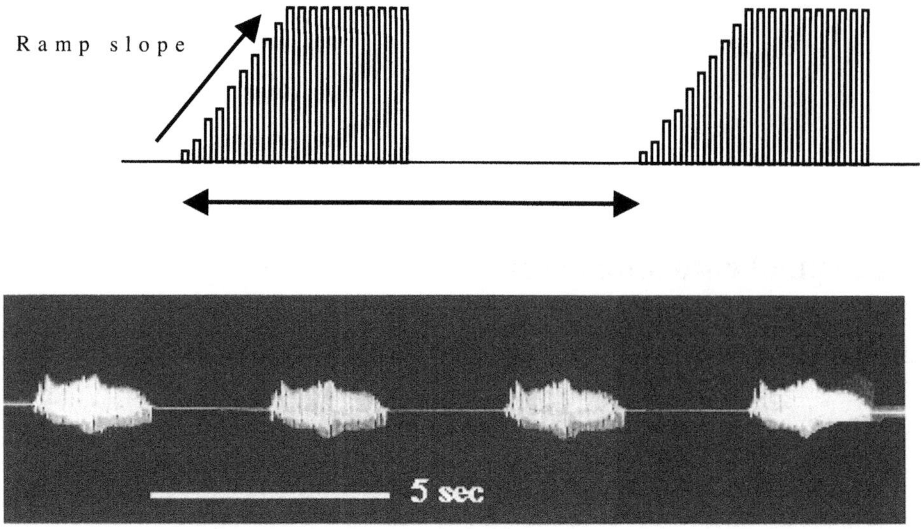

Fig. 1. (Upper) Scheme of the cyclic mode; 2 seconds on with 1 second ramp up, and 3 seconds off. (Lower) Waveform of the output of spinal cord stimulator adjusted for diaphragm pacing

to the reports on optimal stimulation parameters for diaphragm pacing [4, 5], we tested the various types of stimulators for spinal cord stimulation or deep brain stimulation. We found that Itrel 3 and the X-trel system had a function of cyclic output suitable for artificial respiration. The waveforms were checked with an oscilloscope. Based on this study, we decided to set the parameters at the cyclic mode, 2 seconds on with 1 second ramp up, and 3 seconds off, resulting in the respiratory rate being 12 per minute. The pulse width and the frequency were set at 150 μsec and 21 Hz respectively (Fig. 1).

Operative technique

All patients were operated on under general anesthesia without muscle relaxant. A liner horizontal skin incision of 4 cm length was made at 4 cm rostral to the clavicle, crossing the posterior border of the sternocleidomastoid muscle. The phrenic nerve was identified over the scalene muscle and confirmed with monopolar electrical stimulation (5 or 50 Hz, 1 msec pulse width, 0.5–2 volts). Under an operation microscope, the nerve sheath was carefully exposed for about 1 cm in length. A quadripolar lead type electrode for spinal cord stimulation (Medtronic, model 3387-28) was placed along the nerve and fixed with two sutures to the surrounding connective tissues (Fig. 2). The stimulation device (Medtronic Itrel 3, model 7425) was implanted subcutaneously in the anterior upper chest in four patients, and the lead was tunneled and connected between the stimulating electrode and the pulse generator. In one patient (Case 2), we used a radiofrequency coupled device (Medtronic X-trel, Model 3470. In case 5, we performed bilateral diaphragm pacing.

At first, we set the output of the stimulator as in the studies in vitro. The amplitude of the output was adjusted so that the tidal volume became approximately 300–500 ml. The most proximal contact of the quadripolar electrode to the stimulator was set as negative and the most distal contact as positive. The others were set as "off". And the amplitude was adjusted for the tidal volume to be approximately from 300 to 500 ml and for the carbon dioxide pressure in the end tidal air to be around 40 mmHg with the anesthesia monitor. After the operation, respiration with phrenic nerve pacing was started for a short period and the stimulating duration was lengthened under monitoring of the carbon dioxide pressure of the exhalation, the arterial oxygen saturation, and the tidal volume. The carbon dioxide pressure in the end tidal air was adjusted to be about 40 mmHg by changing the output voltage of the stimulator. We did not change the respiratory rate except in case 1. In order to avoid fatigue of the phrenic nerve and diaphragm [1, 4, 5], we used the stimulator only at daytime in patients who required artificial ventilation for 24 hours (Case 2–4). As for a patient with sleep apnea and sufficient ventilation in the awake state (Case 1), we trained her to switch on the device before going to bed by herself.

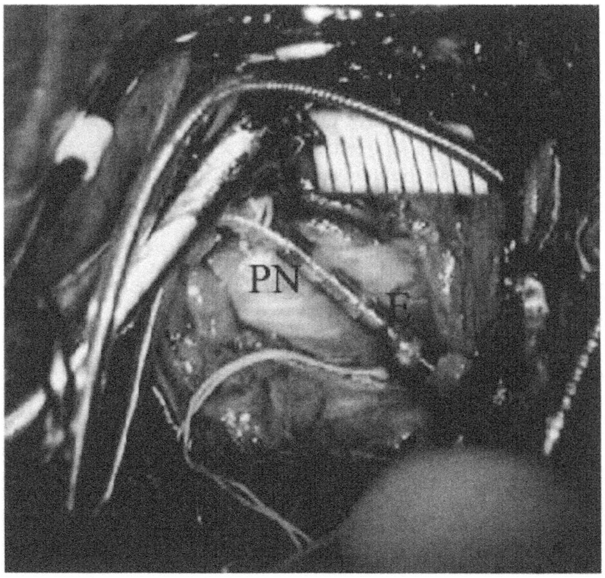

Fig. 2. Photography of electrode placement. *E* Electrode, *PN* phrenic nerve

Table 2. *List of the implanted device, stimulation voltage, and outcome*

Case	Implanted device	Initial setting of stimulation voltage (v)	Final setting of stimulation voltage (v)	Follow-up (months)	Outcome
1	Itrel 3	1.5	2.5	34	free from respirator
2	X-trel	1.2	2.2	30	respirator only at night
3	Itrel 3	1.8	2.8	27	respirator only at night
4	Itrel 3	1.3	2.3	21	respirator only at night
5 (Rt)	Itrel 3	2.5	2.5	3	3–4 hrs respirator free
5 (Lt)	Itrel 3	1.9	1.9	3	

Results

No patient had trouble related to the stimulation system during the follow-up period ranging from 3 to 34 months (mean 23 months). There was no operative complication. Sufficient ventilation was accomplished with the stimulation in all patients after the operation. As summarized in Table 2, the initial stimulation strength determined at surgery was 1.7 ± 0.5 (mean $\pm$ SD) volts. With this strength, the tidal volume became about 500 ml in four cases and 300 ml in Case 5. To maintain carbon dioxide pressure measured in the exhalation at about 40 mmHg, the output strength was gradually increased over several weeks to months. The final setting of the output was 2.4 ± 0.3 (mean $\pm$ SD) volts at the latest follow-up. One patient with sleep apnea syndrome (Case 1) became respirator free and she used the device only during sleep. In another patient (case 5), she is lengthening the period of phrenic nerve pacing. At first, she complained of dyspnea over 20 minutes because of insufficient tidal volume. But respiratory training made her respirator free without dyspnea for several hours. Other patients stopped using a ventilator during daytime. A dash between the artificial and spontaneous breathing did not occur in all patients except one (Case 1). By decreasing the respiration rate to 10 per minute with stimulator setting at 2 seconds on and 4 seconds off, the fighting phenomenon completely disappeared.

Discussion

Artificial ventilation with phrenic nerve stimulation has a long history [2, 4, 5]. Since the device became commercially available (Avery Laboratories Inc., New York), it has been implanted in many patients [1–3]. The efficacy and problems connected with use of this device system have been documented in detail [1–5]. The reason why we started using the spinal cord stimulator for diaphragm pacing is that the device specifically made for phrenic nerve stimulation is not readily available in our country. Contrary to the availability of the device for diaphragm pacing, the spinal cord stimulator is commonly used and covered by insurance in Japan.

The longevity of the battery in Itrel 3 stimulator is about 5 years when used for pain control with continuous output. Compared with the ordinary stimulation parameters for pain control (for instance, 50 Hz, 200 μsec pulse width, 3 volts, and 8 hour continuous stimulation per day), the output parameters for diaphragm pacing in the present cases result in only about 12.5% in terms of electricity consumption. Simple calculation with such information indicates that the battery longevity of Itrel 3 when used for phrenic nerve stimulation is theoretically eight times longer than in spinal cord stimulation for pain control.

We used an electrode for spinal cord stimulation and placed it along the phrenic nerve. Because the electrical current may spread to the surrounding tissues with such an electrode, we had worried that local muscle contraction, electrically evoked sensation, or increased electrical resistance due to surrounding fibrosis might happen. However, we did not experience such problems.

Bilateral implantation of the pacing system is ideal to accomplish full time respiration [2, 3]. However, because we used the device system as a compromise rescue procedure, we did not implant bilaterally in the first four cases. Since the present study showed that the device system had no major problems, we performed bilateral implantation in the last case. The use of spinal stimulator for diaphragm pacing has no warranty by the manufacturer. An emergency backup system is not established. Therefore, we performed the procedure solely on our own responsibility.

Although long-term follow-up is necessary, we conclude that the spinal cord stimulator system is applicable to phrenic nerve stimulation for diaphragm pacing as a compromise.

References

1. Chervin RD, Guilleminault C (1997) Diaphragm pacing for respiratory insufficiency. J Clin Neurophysiol 14: 369–377
2. Elefteriades JA, Quin JA (1998) Diaphragm pacing. Chest Surg Clin N Am 8: 331–357
3. Garrido-Garcia H, Mazaira Alvarez J, Martin Escribano P, Romero Ganuza J, La Banda F, Gambarrutta C, Garcia ME, Labarta C, Arroyo O, Sebastian de la Cruz F, Gutierrez R, Garcia Moreno J (1998) Treatment of chronic ventilatory failure using a diaphragmatic pacemaker. Spinal Cord 36: 310–314
4. Glenn WW, Holcomb WG, Shaw RK, Hogan JF, Holschuh KR (1976) Long-term ventilatory support by diaphragm pacing in quadriplegia. Ann Surg 183: 566–577
5. Glenn WW, Phelps ML (1985) Diaphragm pacing by electrical stimulation of the phrenic nerve. Neurosurgery 17: 974–984

Correspondence: Tetsuryu Mitsuyama, M.D., Department of Neurosurgery, Neurological Institute, Tokyo Women's Medical University, 8-1 Kawada-cho, Shinjuku-ku, Tokyo 162-8666, Japan. E-mail: ttaira@nij.twmu.ac.jp

Neurosurgical intervention for functional recovery from neurological deficit: Part 2

Acta Neurochir (2003) [Suppl] 87: 95–97

Combined dorsal root entry zone lesions and neural reconstruction for early rehabilitation of brachial plexus avulsion injury

H. J. Chen[1], **K. Lu**[1], and **M. C. Yeh**[2]

[1] Department of Neurosurgery, Chang Gung University and Medical Center at Kaohsiung, Kaohsiung Hsien, Taiwan
[2] Department of Plastic Surgery, Chang Gung University and Medical Center at Kaohsiung, Kaohsiung Hsien, Taiwan

Summary

Brachial plexus avulsion injury is one of the major complications after traffic, especially motorcycle accidents. During the past 12 years, we have encountered more than 40 brachial plexus avulsion injuries. The neurological deficits included pain and paralysis of the damaged limb. Dorsal root entry zone lesions made by thermocoagulation were performed for intractable pain in 34 cases. The pain relief rate was good in about 75%. Combined neural reconstruction was performed in 15 cases. The reconstruction included neurolysis, nerve graft, nerve transfer, and functioning muscle/tendon transfer etc. There were 13 male and 2 female patients. Age distribution was from 21 to 61 years with a mean age of 41.8 years. Eleven patients were found to have whole brachial plexus injury and 4 with upper brachial plexus injury. Twelve patients had good pain relief. Six patients showed good functional result after reconstruction. Three had no improvement. Combined pain control and reconstruction offer an early rehabilitation for brachial plexus avulsion injury.

Keywords: Brachial plexus avulsion injury; dorsal root entry zone; neural reconstruction; rehabilitation.

Introduction

Brachial plexus avulsion injury is one of the major complications after traffic, especially, motorcycle accidents [2, 3]. Motorcycles are the main transportation vehicle in developing countries. Consequences of brachial plexus injury may be intractable pain, especially pain after root avulsion and paralysis. Since the majority of patients are young, it is a profound implication if employment and social opportunities are limited by their disability. Early management and subsequent reconstruction will be allowed after appreciation of brachial plexus injury [3, 6].

Pain occurs frequently after injury, starting within days or weeks of the event and usually becomes chronic. Medical management often remains with poor response. Surgical treatment including dorsal root entry zone lesions have been used for pain control [2, 7, 9, 10]. For brachial plexus palsy, neurolysis, neurotization and functioning muscle/tendon transfer have been performed by neurosurgeons and plastic surgeons for restoration of the upper limb function [3, 4–6, 8]. Over the last 12 years, a number of patients with a brachial plexus injury have been treated for reconstruction and pain control. We present our experiences with this kind of injury.

Materials and methods

Between 1990 and 2001, we treated more than 40 cases with brachial plexus avulsion injury. The main complaints were intractable pain and paralysis of the involved targets. Injuries were evaluated by physical examination, electrodiagnostic studies and imaging studies including myelography, computed tomography or magnetic resonance imaging occasionally [1].

Thirty-four patients underwent dorsal root entry lesions with thermocoagulation for pain control. Among these 34 cases, combined reconstruction for brachial plexus injury was performed in 15 cases.

There were 13 male and 2 female patients. Age distribution was from 21 to 61 years with a mean age of 41.8 years. Nine patients had lesion on the left side and 6 patients on the right. Injuries due to motorcycle accidents were found in 13 patients, due to car accidents in one patient, and machine injury in the remaining one.

All patients were found to have poor response to medical treatment and underwent dorsal root entry zone (DREZ) lesions by thermocoagulation for pain control. DREZ surgery may be performed before or after nerve reconstruction, depending on the clinical course of the disease. The surgical principles were according to the Nashold's method [9, 10]. The hemilaminectomy areas were dependent on the avulsed roots. For a complete brachial plexus avulsion injury, hemilaminectomy from C_4 to T_1 was performed to expose the whole lesion site (Fig. 1). The electrode had an internal thermistor to measure the temperature of the lesion and was in-

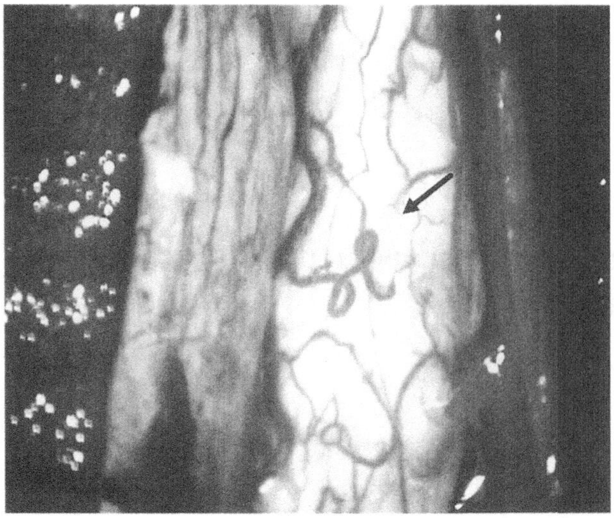

Fig. 1. All dorsal roots were torn (arrow) after opening the dura

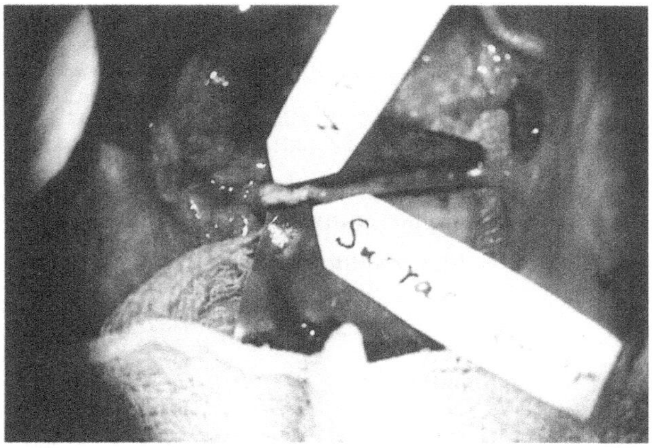

Fig. 2. The spinal accessory nerve is transferred to the suprascapular nerve with end to end anastomosis

troduced into the cord along the posterior lateral sulcus into the DREZ. Each lesion was made at a temperature of 75 °C for 15 to 20 seconds. The interval between two lesions was about 1 mm.

Before reconstruction of brachial plexus injury, we took a detailed record of the change of neurological conditions in a 3-month period. If it was indicated, stages of reconstruction were performed. The procedures included nerve graft for ruptured roots, trunks, cords and nerves, nerve transfer for avulsed roots which included spinal accessory, intercostals and phrenic nerves (Fig. 2) and functioning muscle/tendon transfer for late cases. Some adjuvant procedures such as arthrodesis, osteotomy and tendon transfer were performed at a later stage.

It was necessary to keep immobilization for nerve repair or functioning muscle transfer for a period of one to two months. Rehabilitation procedures would be performed within one month after surgery. Adjuvant procedures could be done even years later. This could help patients for better target organ function.

Results

Twelve patients had good initial pain control after DREZ surgery. Three patients had only fair results. After longterm follow-up, some recurrence of pain was found but could be controlled with medical treatment. All 15 patients underwent different procedures for neural reconstruction, depending on the individual deficits. Detailed information is shown in Table 1.

Discussion

The clinical manifestations after brachial plexus avulsion injury are mainly intractable pain and paralysis of the involved target organs. The goal of treatment is early return to social activities or even work [3, 6]. The deafferentation pain after root avulsion is very incapacitating and several of our patients had suicide histories [2]. In our experience, pain may happen in about one fourth of the patients and good pain control usually is the priority of treatment. DREZ surgery offers an acceptable result in pain control [2, 7, 9].

Reconstruction of the brachial plexus never fully restores the function of the paralyzed limb to normal. Advances in nerve reconstruction surgery have been based on a thorough understanding of the pathophysiology of nerve injury. Neurolysis, nerve graft, nerve transfer and functioning muscle/tendon transfer afford concepts that are essential in understanding the rationale behind the management of these injuries [3–6].

For pain control, our experience suggests that DREZ surgery can be performed as early as possible. There were no major complications encountered in our 34 cases. Although opinions differ regarding the timing of exploration, most authors recommend surgery within 3 to 6 months of injury, thus minimizing muscle atrophy and fibrosis. Most patients in our series underwent exploration within this period. We feel that time for observation is essential in order to find appropriate treatment for each patient. Of course, some adjuvant procedures such as arthrodesis, osteotomy and tendon transfer can be performed even years later.

In conclusion, treatment of brachial plexus avulsion injury is team work. The team includes neurosurgeons, plastic surgeons, hand surgeons, orthopedic doctors, rehabilitation group and psychiatrists. Good pain control and appropriate neural reconstruction offer a better rehabilitation work.

Table 1. *Clinical information of patients who underwent DREZ surgery and brachial plexus reconstruction*

Case No.	Age (years)/Sex	Brachial plexus injury	Follow-up (years)	Result after DREZ surgery	Procedures for reconstruction	Outcome of target
1	61/M	whole	10	good	intercostal nerve and sural nerve graft to the axillary nerve.	fair
2	43/M	whole	12	fair	flexoplasty of the elbow. Transfer of the common flexor tendon to the humerus.	fair
3	55/M	upper	11	good	sural nerve graft, intercostal nerve to the axillary nerve, gracilis for the biceps.	good in flexion of elbow
4	21/M	whole	9	good	sural nerve graft and intercostal nerve transfer to the musculocutaneous nerve.	fair in biceps function
5	51/M	upper	7	good	latissimus dorsi flap to biceps.	fair
6	36/F	whole	5	good	intercostal nerve transfer to the musculocutaneous nerve.	good
7	34/F	whole	8	good	intercostal nerve transfer to the musculocutaneous nerve.	poor
8	35/M	whole	5	good	intercostal nerve transfer to the musculocutaneous nerve.	fair
9	37/M	whole	4	good	intercostal nerve transfer, elbow joint disarticulation.	good in elbow flexion
10	57/M	whole	3	fair	intercostal nerve transfer.	poor in elbow flexion
11	38/M	whole	7	good	neurolysis only.	poor
12	31/M	whole	3	good	spinal accessory nerve to the axillary nerve.	good in shoulder abduction
13	51/M	whole	10	fair	1. spinal accessory nerve to the suprascapular nerve. 2. T4–5 intercostal nerve to the musculocutaneous nerve. 3. T3–4 intercostal nerve to the axillary nerve.	good
14	37/M	upper	10	good	intercostal nerve transfer to the musculocutaneous nerve.	fair in elbow flexion
15	41/F	upper	6	good	sural nerve graft to the lateral and posterior cords, gracilis functioning muscle flap for the biceps.	good in elbow flexion and wrist extension

References

1. Alon M, Rochkind S (2002) Pre-, intra-, and postoperative electrophysiologic analysis of the recovery of old injuries of the peripheral nerve and brachial plexus after microsurgical management. J Reconstr Microsurg 18: 77–82
2. Chen HJ (1992) Dorsal root entry zone lesions in the treatment of pain following brachial plexus avulsion and herpes zoster. J Formos Med Assoc 91: 508–512
3. Chuang DC, Epstein MD, Yeh MC (1993) Functional restoration of elbow flexion in brachial plexus injuries: results in 167 patients. J Hand Surg-Am 18(2): 285–291
4. Chuang DC (1995) Neurotization procedures for brachial plexus injuries. Hand Clin 11(4): 633–645
5. Chuang DC (1995) Functioning free muscle transplantation for brachial plexus injury. Clin Orthoped Related Res 314: 104–111
6. Fogarty BJ, Brennen MD (2002) Upper root brachial plexus trauma; patient selection and reconstruction. Injury 33(1): 57–62
7. Ishijima B, Shimoji K, Shimizu H (1988) Lesions of spinal and trigeminal dorsal root entry zone for deafferentation pain: experience of 35 cases. Appl Neutophysiol 51: 175–187
8. Luedemann W, Hamm M, Blomer U (2002) Brachial plexus neurotization with donor phrenic nerves and its effect on pulmonary function. J Neurosurg 96(3): 523–526
9. Nashold BS Jr (1988) Neurosurgical technique of the dorsal root entry zone operation. Appl Neurophysiol 51: 136–145
10. Sindou M, Daher A (1988) Spinal cord ablation procedures for pain. In: Dubner A, Gebbart GF, Bond MR (eds) Proceedings of the Fifth World Congress on pain. Elsevier, Amsterdam, pp 477–495

Correspondence: Han-Jung Chen, M.D., Ph.D., Department of Neurosurgery, Chang Gung Memorial Hospital, 123, Ta-Pei Road, Niao-Sung Hsiang, Kaohsiung Hsien, Taiwan. E-mail: chenmd@ms8.hinet.net

Acta Neurochir (2003) [Suppl] 87: 99–102

Functional posterior rhizotomy for severely disabled children with mixed type cerebral palsy

N. Morota[1], **S. Kameyama**[2], **M. Masuda**[2], **M. Oishi**[2], **A. Aguni**[3], **T. Uehara**[3], and **K. Nagamine**[3]

[1] Department of Neurosurgery, National Center for Child Health and Development, Tokyo, Japan
[2] Epilepsy & Functional Neurosurgery, National Nishi-Niigata Central Hospital, Niigata, Japan
[3] Department of Orthopedic Surgery, Okinawa Preferectural Naha Hospital, Okinawa, Japan

Summary

The authors evaluated the impact of functional posterior rhizotomy (FPR) for children with severely disabled mixed type cerebral palsy (CP).

Three quadriplegic children at the age of 3, 4, and 10 years underwent FPR. They were classified as mixed type CP based on the clinical presentation of marked spasticity with dystonic posture. Preoperative Ashworth score of the lower extremity was 3.5, 4.5, 4.8 respectively. Two children showed prominent opisthotonus and all showed severe subluxation of the hip joint. Advanced scoliosis was associated in two children. FPR was performed from L2 to S1 in one child, L2 to S2 in one and L2 to S1/S2 in one based on the result of pudendal mapping. Rootlet cutting rate ranged from 66 to 75%. Postoperatively, Ashworth score dropped to 1.4, 1.2, 1.3, respectively. Functional improvement of the upper extremity and urination were confirmed in two children. Hip subluxation was reduced in one child and remained stable in two. A one-year follow-up review confirmed no relapse of spasticity among them.

FPR achieved highly satisfactory surgical effects in children with severe mixed type CP. Although long-term follow-up is mandatory since there was a report of relapsed spasticity after FPR in this particular population of CP, FPR could be a choice of surgery in severely disabled children with mixed type CP.

Keywords: Rhizotomy; cerebral palsy; spasticity; dystonia.

Introduction

Functional posterior rhizotomy (FPR) is performed to reduce spasticity for children with spastic cerebral palsy (CP) [1, 4, 11]. Modern FPR emerged as a new neurosurgical procedure for spastic CP in the late 1970s [6]. The author (NM) introduced the procedure in 1996 but it still remains relatively new in Japan. Recently, great attention has been paid to FPR and the horizon now seems to gradually open for the treatment of spastic CP.

The role of FPR for severely disabled children with CP has been unclear despite the widely recognized fact that in general it marked reduction of spasticity for children with spastic CP [1, 7, 9–11]. It is not unusual that a severe form of spastic CP can be associated with other forms of involuntary movement like dystonia. Although association of athetosis often precludes surgical indication of FPR [10], the role of FPR for this particular group of children needs further to be discussed.

In this paper, the author reviewed surgical results of 3 children who had severely disabling spasticity associated with dystonia.

Materials

Ten children with spastic CP underwent FPR from March 1996 to May 2002. Among them, 3 quadriplegic children aged 3, 4, and 10 years, who were ranked as grade 5 in the New York University Medical Center patient group, were diagnosed to be of mixed type of CP since dystonic posture was confirmed in their motion [2]. They had a totally dependant life and they were unable to roll over or move by themselves. Daily care for them was a great burden for the care-givers because of severe spasticity. Causes of CP were either severe asphyxia or hypoxia in the perinatal period. Two children showed marked ventriculomegaly on CT and diffuse cerebral hypoperfusion on SPECT. Severe scoliosis (Cobb's angle > 40°) was associated in two children. Follow-up period was 1 year and 3 months in one, and 1 year in two children.

Methods

FPR after L1–L5 osteoplastic laminotomy and S1–S3 laminectomy was performed under general anesthesia using intravenous infusion of Propofol. Range of lesioning was decided based on the result of pudendal mapping [5]. One child underwent FPR from bilateral L2 to S1, one from bilateral L2 to S2 and the last one from L2 to S1 in one side and the other from L2 to S2. Rootlets to be cut were judged by muscle responses on EMG following electrical stimula-

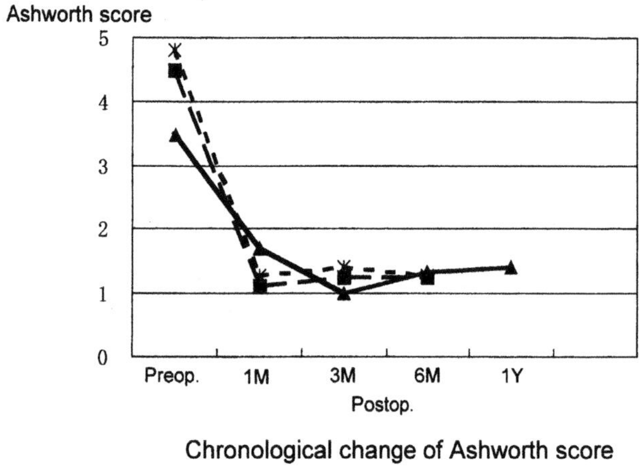

Fig. 1. Chronological change of Ashworth score showed marked reduction of spasticity immediately after FPR. The result of surgery has been maintained during the follow-up period in all children

tion. Those who showed highly abnormal responses were cut after coagulation.

Results of surgery, chronological change of Ashworth score and the score of GMFM (Gross Motor Function Measurement), and other factors influencing the daily activities were evaluated.

Results

FPR was safely performed even in the two patients who had severe scoliosis. Cutting rate of the dorsal rootlets was 66%, 71%, and 75%. The cutting rate was higher than in other children who had milder forms of spasticity without dystonia. Only one child complained of worsened constipation but no other significant complication was noticed or reported by the family.

Preoperative average Ashworth score of the lower extremity was 3.5, 4.5 and 4.8, which, after FPR, dropped to 1.4, 1.2 and 1.3 respectively. The initial surgical effect has been maintained up to date (Fig. 1). GMFM showed mild improvement in one patient but in the other two whose preoperative score was zero remained the same postoperatively.

Prominent opisthotonus improved in all, but one child showed mild relapse of opisthotonic posture when hold in hands. Functional improvement of the upper extremity (smooth, more relaxed movement) and urination (longer dry time) was noticed in two children. Subluxation of the hip joint was reduced in one child and remained stable in two [8].

Family satisfaction following FPR was reported in all. They confirmed a marked reduction of daily care burden after FPR (Fig. 2).

Discussion

FPR seemed to be effective for children with severely disabled mixed type CP. It has been reported that reduction of spasticity following FPR is most effective in children with spastic CP [10]. Surgical result of FPR for spasticity associated with athetoid CP was less encouraging and it is not regarded as surgical indication. Although our experience is quite limited in number, it clearly demonstrated that FPR is not only effective to reduce spasticity associated with dystonia but also useful for improving abnormal posture or muscle contraction in dystonia. Longer follow-up of the effect of FPR in this particular group of children is mandatory since relapse of spasticity could develop later on. Fasano reported from his experience as a pioneer in the modern FPR that relapse of spasticity was confirmed in one fourth of children with severe form of spastic CP, although association of dystonia in the patients and cutting rate of the dorsal rootlet were not mentioned in the paper [7].

High cutting rate of the rootlet didn't associate with increased rate of complications. It is natural to speculate that higher rate of rootlet cutting will bring better results of controlling spasticity at the expense of more functional loss. However, facts taught us that this assumption was not true and the patient could tolerate the higher rate of rootlet cutting with nearly no complication in terms of sensory and motor function. It may depend on the preexisting severe brain damage which conceals the functional disturbance after the surgery. In addition, goal of FPR for these severely disabled children is to decrease the burden of care givers, not to gain any functional improvement of the lower extremity. These factors support higher cutting rate of the rootlets in severely disable children with CP when indicated by the intraoperative neurophysiological examination. More time and longer follow-up are mandatory before reaching a final conclusion.

Controversy stands for surgical indication of FPR and intrathecal Baclofen (ITB) therapy [12]. FPR will remain an important choice of surgery even after clinical introduction of ITB therapy for severely disabled children with CP [3]. FPR is a single procedure and surgical results could be effective for a long time. On the other hand, ITB therapy is less invasive in surgery but periodic refilling of Baclofen is necessary. A higher daily dosage of Baclofen is said to be required for severely disabled mixed type CP, patient which means higher expenses in the long term. Risk of infection and

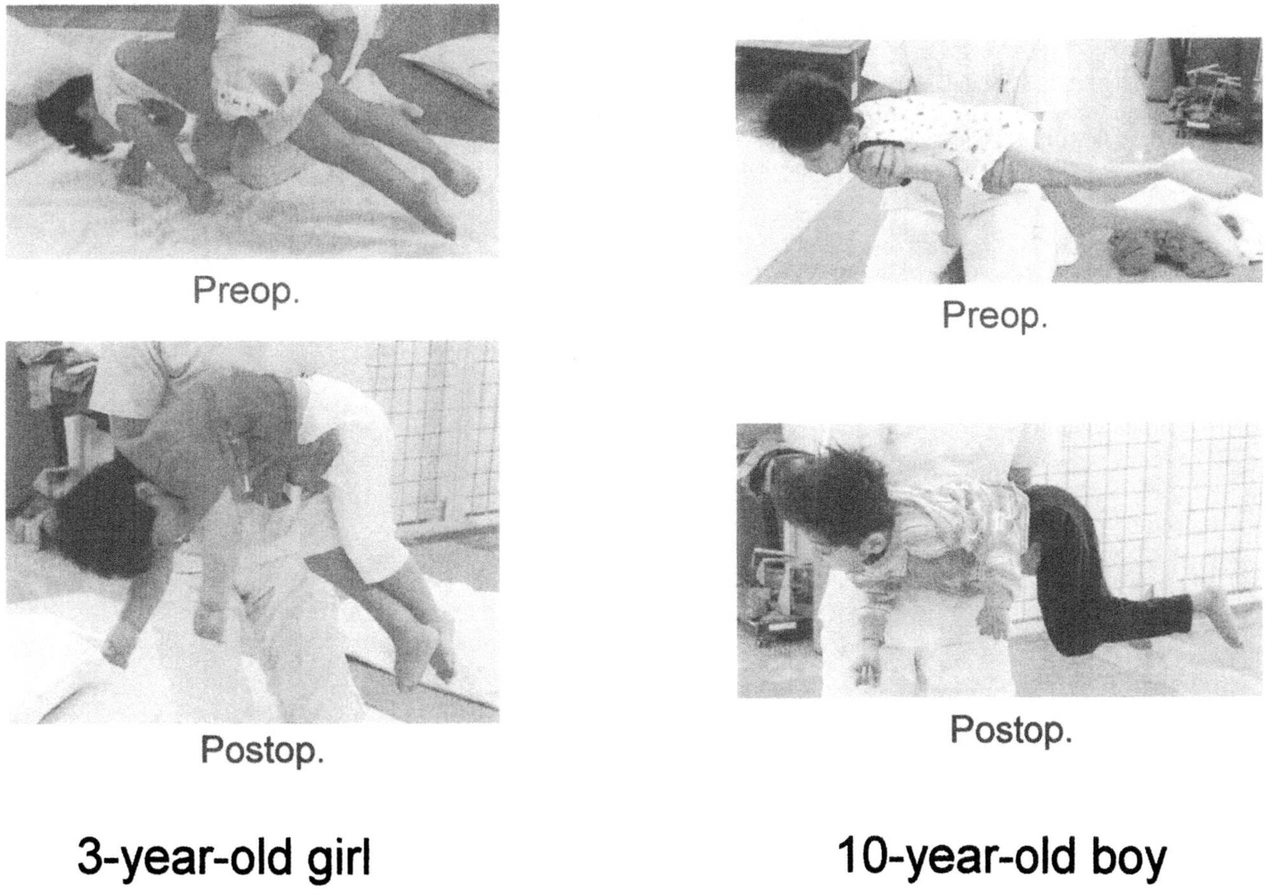

Fig. 2. Pre and postoperative change of the patient's posture is showed in two children. Reduced spasticity and improved dystonic posture are demonstrated while the children are supported by hands in the prone position

financial burden could be a problem especially when the patient is institutionalized. Although possible recurrence of spasticity could be a major drawback of FPR, it seems to be worthwhile performing as a first step of surgical treatment for this particular patient group [7].

Summary

FPR achieved highly satisfactory surgical results in children with severely disabled children with dystonia. Although long-term follow-up is mandatory since relapse of spasticity after FPR could not be excluded in this particular population of CP, it is suggested that FPR could be a choice of surgery in cases of severely disabled mixed type CP.

References

1. Abbott R, Forem SL, Johann M (1989) Selective posterior rhizotomy for the treatment of spasticity: a review. Childs Nerv Syst 5: 337–346
2. Abbott R (1991) Indication for surgery to treat children with spasticity due to cerebral palsy. In: Sindou M, Abbott R, Keravel Y (eds) Neurosurgery for spasticity. A multidisciplinary approach. Springer, Berlin Heidelberg New York Tokyo, pp 215–217
3. Albright AL, Cervi A, Singletary J (1991) Intrathecal baclofen for spasticity in cerebral palsy. JAMA 265: 1418–1422
4. Arens LJ, Peacock WJ, Peter J (1989) Selective posterior rhizotomy: a long-term follow-up study. Childs Nerv Syst 5: 148–152
5. Deletis V, Vodusek D, Abbott R, Epstein F, Turndorff H (1992) Intraoperative monitoring of the dorsal roots: minimizing the risk of iatrogenic micturation disorders. Neursurgery 30: 72–75
6. Fasano VA, Urciuoli R, Broggi G, Barolat-Romana G, Benech F, Ivaldi A, Sguazzi A (1977) New aspects in the surgical treatment of cerebral palsy. Acta Neurochir (Wien) [Suppl] 24: 53–57
7. Fasano VA, Broggi G (1989) Functonal Posterior Rhizotomy. In: Park TS (ed) Neurosurgery: state of the art reviews 4. Management of spasticity in cerebral palsy and spinal cord injury. Philadelphia, Hanley & Belfus, pp 409–412
8. Heim RC, Park TS, Vogler GP, Kaufman BA, Noetzel MJ, Ortman MR (1995) Changes in hip migration after selective dorsal rhizotomy for spastic quadriplegia in cerebral palsy. J Neurosurg 82: 567–571
9. Morota N, Abbott R, Kofler M, Epstein FJ, Cohen H (1995) Residual spasticity after selective posterior rhizotomy. Childs Nerv Syst 11: 161–165

10. Peacock WJ, Arens LJ (1982) Selective posterior rhizotomy for the relief of spasticity in cerebral palsy. S Afr Med J 62: 119–124
11. Steinbok P (2001) Outcomes after selective dorsal rhizotomy for spastic cerebral palsy. Childs Nerv Syst 17: 1–18
12. von Koch CS, Park TS, Steinbok P, Smyth M, Peacock WJ (2001) Selective posterior rhizotomy and intrathecal baclofen for treatment of spasticity. Pediatr Neurosurg 35: 57–65

Correspondence: Nobuhito Morota, M.D., Department of Neurosurgery, National Center for Child Health and Development, 2-10-1 Ohkura, Setagaya-Ku, Tokyo 157-8535, Japan. E-mail: morota-n@ncchd.go.jp

Acta Neurochir (2003) [Suppl] 87: 103–105

The role of neurosurgical interventions for control of spasticity in neurorehabilitation: new findings on functional microanatomy of the tibial nerve

T. Taira and **T. Hori**

Department of Neurosurgery, Neurological Institute, Tokyo Women's Medical University, Tokyo, Japan

Summary

Harmful spasticity after cerebral or spinal damage disturbs functional recovery in neurorehabilitation, but neurosurgical interventions for relief of spasticity are not widely performed, at least in Asian countries including Japan. We have been performing various types of neurosurgical treatment for spasticity such as selective peripheral neurotomy, selective dorsal rhizotomy, microsurgical DREZotomy, and intrathecal baclofen administration. We deal with both children and adults. From our experience of tibial neurotomy, in particular, we describe functional microsurgical anatomy of the tibial nerve in the popliteal fossa. This is the first report on this regional functional anatomy.

Keywords: Spasticity; neurosurgical intervention; tibial nerve.

Introduction

Neurosurgical interventions for harmful spasticity play an important role to improve efficacy of rehabilitation after cerebral or spinal damage. However, the type of neurosurgical treatment for spasticity varies from country to country. For example, selective peripheral neurotomy is seldom performed in the United States, while it is a common procedure in Europe, especially in France. Botulinum toxin is used for temporal control of focal spasticity in the US, while it is not approved in Japan. Every neurosurgical procedure for control of spasticity has pros and cons, and its indication should be partly decided according to the socio-economic situation of every country. We have been seeking the appropriate neurosurgical interventions and came to the conclusion that selective peripheral neurotomy is a reasonable treatment for focal spasticity in mild cerebral palsy or after stroke. We would like to show some new findings from our experience of selective peripheral denervation.

Materials and methods

During the past 3 years, we have performed selective peripheral denervation for control of focal spasticity in 110 patients. Seventy-eight patients were cerebral palsy and 70 received selective neurotomy of the tibial nerve for equines and/or varus spasticity of the foot. Forty-five of them underwent bilateral procedures. Although the procedure on tibial neurotomy has been well described by Sindou and Mertens [6, 7], we noticed that there have been no reports on functional microsurgical anatomy of the tibial nerve in the popliteal fossa.

Results

The senior author (T. T.) carried out the operation from skin incision to closure. The patient was placed in the prone position under general anesthesia without muscle relaxant. The skin incision in the popliteal fossa is as shown in Fig. 1 and it is about 2.5 cm in length. We used fine monopolar electrode (tip diameter: 0.2 mm) for identification of motor nerves. The wound retractor was used as the reference electrode. The parameters of the stimulation were 0.2 msec pulse and square electrical wave form. We used both 5 and 50 Hz stimulation. The strength was adjusted at 0.2–3 volts according to the strength of the muscle contractions. We defined motor nerve as when the muscle contraction is elicited with stimulation. When stimulation showed no muscle response, we regarded the nerve as sensory. From our experience of tibial neurotomy, we came to a general conclusion as follows. The findings are schematically shown in Fig. 1.

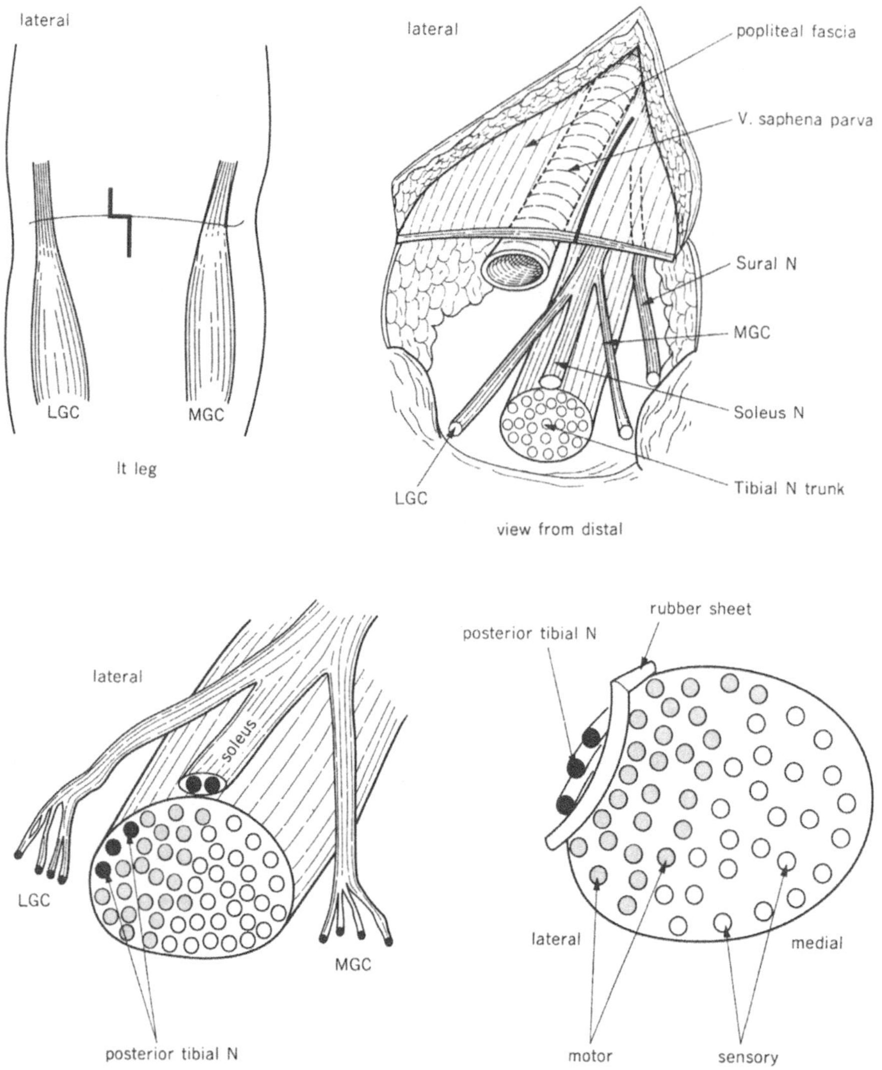

Fig. 1. Functional microsurgical anatomy of the tibial nerve and its branches at the popliteal fossa. *LCG* Lateral gastrocunemius nerve, *MGC* medial gastrocunemius nerve

1. After skin incision and dissection of subcutaneous fat tissues, a small vein (V saphena parva) is transparently observed under the popliteal fascia. The fascia should be incised longitudinally and medial to the vein. This makes exposure of the tibial nerve trunk easier. (Fig. 1 upper right)
2. The sural nerve (sensory branch of the tibial nerve) runs in the fat tissue under the popliteal fascia, and should be carefully preserved.
3. The soleus nerve and the medial and lateral gastrocnemius nerves have a common trunk in the proximal part and it lies on the dorsal aspect of the tibial nerve trunk. (Fig. 1 upper right, lower left)
4. The soleus nerve has one or two nerve fasciculi. (Fig. 1 lower left)
5. The medial and lateral gastrocnemius nerves are composed of 4–7 small nerve fasciculi. (Fig. 1 lower left)
6. The posterior tibial nerve for varus spasticity lies in the dorsolateral part of the tibial nerve trunk. There are two or three nerve fasciculi for posterior tibial nerve in this region. (Fig. 1 lower right)
7. In the tibial nerve trunk, motor nerves lie laterally and sensory nerves medially. (Fig. 1 lower right)

Discussion and conclusion

There have been many reports on tibial neurotomy [1–7] and its effects on spastic foot such as equines and varus have been well established. Although there is higher tendency of recurrence in children with cerebral palsy [2], the procedure is still important in controlling the focal spasticity of the foot. It is surprising that there have been no descriptions on functional and surgical microanatomy of this nerve as is presented in this paper. We have seen two patients who complained of dysesthesia of the toes after tibial neurotomy for both equines and toe flexion. When relieving toe flexion spasticity, we dissect the tibial nerve trunk itself and find the nerves responsible for toe flexors. Because the location of the motor nerves in the tibial nerve trunk had remained unknown, excessive dissection may have resulted in such complications. As shown in this paper, the distribution of motor nerves in the tibial nerve trunk is always in the lateral part. The posterior tibial nerve lies in the dorsolateral part of the tibial trunk. We believe new knowledge on such topography is important to avoid complications. In case of soleus nerve, it only has one of two nerve fasciculi, and true hyponeurotization is necessary. However, because the medial and lateral gastrocnenius nerves have 4–7 fasciculi, simple resection of some of them is enough to relief spasticity of these muscles.

References

1. Berard C, Sindou M, Berard J, Carrier H (1998) Selective neurotomy of the tibial nerve in the spastic hemiplegic child: an explanation of the recurrence. J Pediatr Orthop B 7: 66–70
2. Caillet F, Mertens P, Rabaseda S, Boisson D (1998) The development of gait in the hemiplegic patient after selective tibial neurotomy. Neurochirurgie 44: 183–191
3. Decq P, Filipetti P, Cubillos A, Slavov V, Lefaucheur JP, Nguyen JP (2000) Soleus neurotomy for treatment of the spastic equinus foot. Neurosurgery 47: 1154–1160
4. Deltombe T, Gustin T, Laloux P, De Cloedt P, De Wispelaere JF, Hanson P (2001) Selective fascicular neurotomy for spastic equinovarus foot deformity in cerebral palsy children. Acta Orthop Belg 67: 1–5
5. Feve A, Decq P, Filipetti P, Verroust J, Harf A, N'Guyen JP, Keravel Y (1997) Physiological effects of selective tibial neurotomy on lower limb spasticity. J Neurol Neurosurg Psychiatr 63: 575–578
6. Sindou M, Mertens P (1988) Selective neurotomy of the tibial nerve for treatment of spastic foot. Neurosurgery 23: 738–744
7. Sindou M, Mertens P (1991) Selective peripheral denervation for the treatment of spasticity. In: Sindou M, Abbott R, Keravel (eds) Neurosurgery for spasticity. Springer, Wien New York, pp 120–132

Correspondence: Takaomi Taira, M.D., Ph.D., Department of Neurosurgery, Neurological Institute, Tokyo Women's Medical University, 8-1 Kawada-cho, Shinjuku-ku, Tokyo 162-8666, Japan. E-mail: ttaira@nij.twmu.ac.jp

Acta Neurochir (2003) [Suppl] 87: 107–112

Restoration of locomotion in paraplegics with aid of autologous bypass grafts for direct neurotisation of muscles by upper motor neurons – the future: surgery of the spinal cord?

K. R. H. von Wild[1] and **G. A. Brunelli**[2]

[1] Medical Faculty of the Westphälische Wilhelms University Münster, Germany
[2] Italy Foundation for Research on Spinal Cord Lesion Brescia, Italy

Summary

Objectives. Paraplegia means a lifelong sentence of sensory loss, paralysis and dependence. Complete spinal cord lesions cannot heal up to now despite intensive experimental research, remarkable efforts and recent achievements in bio-technology and re-engineering. Traumatic paraplegia due to spinal cord injury (SCI) is a quite frequent condition and related to the socio-economical situation of the population. It is experienced disproportionately by young people. The rise in gunshot wounds is dramatic. SCI has appeared refractory to treatment.

Patients and methods. Since 1980 G.A.B. had tried surgical repair of the spinal cord (SC) after experimental bisection in rats, and since 1993 research was done on monkeys (macaca fascicularis) to be closer to human physiology. The sciatic nerve was removed and used as an autologous graft from the lateral bundle of the spinal cord (tractus corticospinalis ventro lateralis) to the three muscles of both legs being known to be most important for locomotion: M. gluteus maximus, M. gluteus medius and M. quadriceps femoris. The first fruitful transplantation in a human being was performed in July 2000.

Results. The results in rats were promising and fulfilled the requirements of the American Task Force of the National Institute of Neurological and Communicative Disorders and Stroke of the US. The results in monkeys confirmed the paradigm so that we performed the first operation in a young lady suffering for four months from complete SC lesion T9 after approval by the ethical committee. First voluntary movements of the connected muscles after 17 months. 27 months after op she was able to walk up to 60 steps with the help of a walker and to climb steps in the water. Improvement is still continuing.

Discussion. SCI has appeared refractory to any kind of treatment. Compensatory strategies are still experimental in human beings. Autologous nerve grafts from the spinal cord tissue (the lateral spinal bundle) connected to peripheral muscle nerves seem promising in paraplegics. But the physiology is still unclear when the glutamatergic upper motor neuron connected to motor end-plates (cholinergic) does work like in our patient.

Conclusion. Further studies in primates and paraplegic patients are necessary to clarify the bypass grafting of the SC to muscle groups distal to the complete SCI to restore locomotion.

Keywords: Spinal cord injury; upper motor neuron grafting: surgical re-engineering in SCI; locomotion in paraplegics; functional neuro-rehabilitation.

Introduction

In Europe there are about 300.000 paraplegics and in every country approximately between 1000 and 1800 new cases per year with an increasing number of tetraplegics. From major reports on the incidence of SCI one may expect between about 30 to 50 victims per million population.

The causes are varied yet, however, they were reported to be consistent over the last decades, related to the given social economical situation of the population. Motor vehicle accidents including motorcycles and bicycles remain in 50% the leading cause, followed by falls and interpersonal violence (gunshot wounds and other assaults) with 15–20% each, followed by sports and recreation [1, 10, 16, 18]. One has to notice the most dramatic exponential rise in SCI because of gunshot wounds in regions and countries of social-economic conflicts e.g. in Brazil as recently reported because of an actual survey by, Marcos Massini, delegate of the WFNS Committee on functional neurosurgical rehabilitation. Sixty percent of all SCI are injured before age 30. More than 90% of the victims may survive in the developed countries due to the quality of both emergency and intensive functional neurorehabilitation with nearly normal experience of live when put in a wheelchair for the rest of their life. It is well known that most patients will recover functionally to some extent from SCI over the first year, but no patient ASIA/IMSOP Grade 4 who remained paraplegic for one year will normally recover to some extent from loss of motor and sensory functions. The important distinction between complete and incomplete is crucial for planning both the acute treatment and rehabilitation as early as possible to achieve the patients' best functional result in terms of the new Classification of Function WHO 2001 as well as for predicting the final outcome concerning recovery and restoration of neurological function below the injury level. There is no hope at all for recovery in case of complete ventral dislocation of the vertebra related to the narrow space within the cervical and thoracic spinal column leading to a most severe "guillotine" lesion of the spinal cord. Conscientious and scrupulous examination and careful planning of all kind of medical treatment is mandatory before starting conservative or even more important surgical reconstructive therapy to to restore motor functions in para- and tetraplegics.

The actual aspects of treatment strategies in SCI reported elsewhere [2, 10, 12, 14, 16, 18] underlining the fact that despite promising animal experiments have shown various treatment possibilities for the acute and chronic phase, for clinical application in human being's there is up to now no practical way to restore the spinal cord for restitution of locomotion, including application of bio- and or gen-technology and all other kind of re-engineering. We, therefore, like other surgeons interested in functional rehabilitation of SCI paralysis became intensively involved in this field over the last twenty years [4–9, 11, 15]. Parallel with our attempts of computer added locomotion ind paraplegic SCI victims level T8–11 by implanted neuroprosthesis for functional electrical stimulation (FES) in paraplegic patients [17], senior author G.A.B. continued his clinical but still experimental work on autologous peripheral nerve by-pass grafts in primates by using a multi-interdisciplinary team approach [5–7]. Our report is focussed on the first successful clinical application of *Brunelli's paradigm* for voluntary restoration of locomotion, based on his experimental concept of plastic reconstructive surgery, in a young lady suffering from posttraumatic complete paraplegia, which challenging surgical concept could be the future in permanent complete thoracic SCI lesions T8–T10.

Material and methods

Since 1980 the senior author GAB has been trying surgical repair of the spinal cord after experimental bisection with complete interruption of the axons within the targeted area of the nervous tissue. First experiments were done on rats by resection of 5 mm the spinal ord tissue and connecting the rostral with the caudal stump by means of autologous peripheral nerve grafts. This was performed according to the microsurgical technique of bridging defects without additional tension as it is used for reconstruction of peripheral nerves lesions [1, 4]. The axons from the central motor neurons have been demonstrated to grow out inside and through the grafts from the peripheral nervous system (PNS) but they stopped progressing as soon as they were again in contact with the central nervous system (CNS) environment at the end of the graft. So it was clearly demonstrated that central neurons were able to regenerate after complete disconnection with interruption to a certain extent their axons within the endoneural tubes of peripheral nerve grafts, which contain Schwann cells. However it became quite clear that reaching the CNS this tissue didn't allow the axons to pass through and to progress regeneration towards their original target. GAB therefore followed a new paradigm of surgical by-pass procedure by connecting distally the nerve grafts and the sprouting axons directly to he peripheral nerves, which of course have been preserved. This was done in his labs between 1982 and 1991 using different surgical protocols, which advanced with improving results year by year. The results became promising and for the first time the results obtained fulfilled the requirements of the American Task Force of the National Institute of Neurological and Communicative Disorders and Stroke of the USA, namely: 1st The experimental lesion must cause disconnection of nerve process, 2nd The process of CNS axons must bridge the level of injuring; 3rd The regenerated fibres must make junction contacts; 4th The regenerated fibres must generate post junction responses; 5th Changing in function must derive from regenerated connections.

The muscles which had been connected by peripheral nerve grafts with the spinal cord were trophic and tonic. They responded to electrical stimulation and stimulation of the graft and of the spinal cord proximal to the experimental disconnection and insertion of the graft.

To be closer to human physiology and neuropathology this experimental work has been continued on monkeys (macaca fascicularis) since 1993. The sciatic nerve was removed and used as an autologous graft, which was inserted one or two segments above the level of spinal cord dissection into the lateral bundle of the cord (cortical-spinal bundle) as a by-pass procedure to be connected with to the motor fascicles of the Mm gluteus maximus, gluteus medius, and quadriceps [1, 3–5]. Re-growing axons were concentrated within the peripheral nerves of the three functional most important muscles to restore locomotion. This technique avoided dispersion of re-growing fibres to less important fascicles and muscles. At the same time spontaneous regeneration of the central neuron connected to the lumbar nerve root was excluded because of complete disconnection of the muscles by cutting the muscular nerve fibres before suturing the nerve graft directly to the nerve stump next to the enter zone of the peripheral nerve into the muscle. With permission from the Italian Ministry the experimental work was continued by GAB in his laboratories of the Italian Foundation for Research on Spinal Cord Lesions in Brescia, Italy, with the second, third and forth group of monkeys since 1996. The final examination included assessment of the monkey's capability of walking and grasping. Monkeys were able to use their previous paralysed lower limb and the connected muscles were found not to be atrophic. The upper motor-neuron produce a cytoskeleton, which was much longer than its original one, and built up new motor endplates. Electrical stimulation gave pretty good muscular responses with different latency times according to the different distances between the targeted muscle and the point of stimulation. Specimens of the muscles reconnected with the nerve graft from the intact spinal cord insertion showed features very similar to those of a repaired peripheral nerve after reconstruction. These studies were able to prove that the upper motor neuron (CNS) can innervate a muscle without the intermediate of the lower second motor neuron of the ventral horns. This findings confirmed the earlier results in rats from the years 1982–1985.

After the proposal of a motion for clinical application of *Brunelli's paradigm* on paraplegic Patients to restore locomotion according to a very strict protocol, which was based on the objective findings in primates, as described above, the National Health Service Ethical Committee of Imola, Italy, finally gave the permission to perform the operation in a patient.

Case report

The patient, a lady, 29 years of age, was otherwise healthy and active as a business women when she had a traffic accident with multiple organ lesions and a most severe dislocated fracture of the thoracic spine vertebrae T9 and 10 with immediate complete lesion level T9 because of a "Guillotine like" dissection of the spinal cord and clinical signs and symptoms level T9 ASIA/IMSOP group A. During the following 4 months no change of the complete neurological deficit could be detected despite all medical therapeutically efforts according to the golden standard in SCI treatment and rehabilitation as follows:

Initial decompress laminectomy T9 and T10, opening of the dural sac with demonstration of anatomical disruption of the spinal cord tissue and hemorrhagic contusion of both stumps, dorsal transpedicular stabilisation after removal of space occupying bony fragments and repositioning of the dislocated T8 vertebra, with an additional attachment of autologous bone material. This procedure went well and the postoperative course was uneventful. Immediate begin of specific neurorehabilitation in the special SCI neurorehab centre Montecatone. When this lady heard and was informed about the experimental results that were achieved in monkeys and the novel possibility to do this procedure in humans, well knowing that this procedure would be the first and a really experimental one, she desired strongly to be operated as soon as possible. She was fully aware about the terrible situation to suffer from the burden of a complete SCI lesion, paralysed, without any realistic hope or chance of spontaneous functional recovery at that time. Moreover she wanted to be operated because of her severely increased muscle tonus in the lower limbs and spasticity, raising severe pain in her back and both legs. After intensive discussions with the experts for neurorehabilitation at the SCI centre and very detailed explanation concerning the operative procedure of bypass nerve grafting to restore locomotion with its typical risks including the fact of resection of both sciatic nerves, she was deeply impressed by the expected benefits. So she gave her consent to be operated and not to wait any longer while muscle atrophy would precede despite intensive physical therapy.

On July 26th, 2000, the operation was performed at the Imola hospital. Surgeon in charge GAB with K.v.W., assisted by G. R. Brunelli: Surgical procedure in general anaesthesia at the begin in prone position, exactly according *Brunelli's paradigm* of by-pass grafting the te spinal cord lesion: Simultaneously operating on both sides at first the sciatic nerves were taken out with micro dissection of the peronel nerves and preparation the Nn musculares to the Mm gluteus maximus and medius next to their entry zone, laminectomy T7, openong of the dural sac where the cord was normal and bilateral insertion of the 45 cm long grafts of both nervi peronei into the ventro lateral cortical-spinal bundles on both sides. The grafts were connected with the nervi musculares mm glutei maximus et medius. Closure of the wounds. Than the patient was turned to her backside, so that the third part of the peronel fasacicles could be placed and sutured to the muscular part of the femoral nerve. This part of the femoral nerve was identified by electrical stimulation during microsurgical neurolysis. The postoperative course was uneventful.

The first voluntary movements were expected to occur after two years, according to the regeneration speed and the length of the grafts used for the by-pass. On the contrary already after 17 months postoperatively the first contractions of the musculus quadriceps occurred. Improvement went on in the following months. After 27 months the lady is able to walk 60 to 70 steps with the help of a walker several times a day. In the water she started to climb the steps by her own. Locomotion is improving day by day continuously while she is still under the control of GAB and her physiatrist. The lady is happy about the results so far achieved and she is highly motivated and therefore willing to continue with the her time-consuming physical training programme.

Discussion

The only way to restore voluntary muscle command, sphincter functions for bladder and rectum evacuation control and cutaneous sensibility to localising in space the paralysed limbs is to bridge the gap of the spinal cord lesion by local and systemic medical treatment in the sense of re-engineering [1, 3–9, 11–16]. This however is demonstrated to be still impossible in humans despite great strides that have been performed in animal experiments and in understanding of the basic physiology and pathophysiology of the spinal cord regarding acute traumatic impact and secondary lesions.

The spinal cord like the CNS is non permissive for the progression of regrowing axons. Reasons of this "non permissively" are not clear yet. Experimental work at the KAROLINSKA Hospital, Stockholm, demonstrated the connection of outgrowing axons with postsynaptic motor neurons of the ventral horn with functional contacts of white to grey matter using peripheral nerve grafts, however limited to only 1 mm without functional connection from white to white matter. Richardson *et al.* [15], Bigamini *et al.* [1], and the senior author [5] demonstrated that when putting grafts of peripheral nerves into the CNS tissue the central outgrowing axons were able to pass through the peripheral nervous system (PNS) endoneural tubes, containing SCHWANN cells, but stopped as soon as again in direct contact with CNS environment. Up to now complete cord lesions cannot heal, because sprouting and regeneration by passing through the white tissue of the spinal cord, the CNS, is not possible for regrowing axons [2, 10, 12]. Several hypotheses have been raised and discussed: The lack of substances favouring axon progression (neutrophins, laminins, diffusible attractant molecules for chemotactic appeal), the lack of Schwann cells which myelinate the regrowing axons in the PNS, the presence of inhibiting substances (NOGO substances) which are normally present in the CNS, the presence of glia cells instead of Schwan cells (ologodendrocyts, astrocyts, and microglia) forming a scar, and a lack of substances (NGF Neurophins) coming from diastal target cells. The literature was recently reviewed by A. Mautes *et al.* and W. Young. None of the numerous procedures for treatment in the chronic stage after SCI has reached the level of clinical trials yet, e.g. neuroprotective therapies, recovery-enhancing treatment including suppression of inflammatory reactions, and regenerative, restorative therapies, in particular, the development and availability of new technologies for transplantation of fetal and genetically modified cells to modify the environment and to deliver growth factors selectively at the site of the injury. According to Volker Dietz from the interdisciplinary task force on functional rehabilitation of SCI in CH-Zürich, this group is now ready to start the first clinical trial based on the new bio-technology (personal communication Venice, 2002).

Paralysis is the most frequently cited handicap in SCI, but much more important is the burden of loss of motor and sensory functions in the sacral segments S4–S5 in complete lesions type Grade A of ASIA/IMSOP Loss of pain sensation leads to pressure ulcers or pressure scares, and many victims suffer from spastic disorders and pain. Restoration of walking is known to be not the most avidly sought of paraplegic patients, who are nearly independent with their wheel chair. However, in respect to prevention of secondary lesions restoration of locomotion could help to reduce the local pressure from the sites of the arms, hands and the buttocks and at the same time it would have a positive effect on the intestinal physiology, not to forget the psychological effect to be able again to stand up and to walk after being completely paralysed. This seems to us to be more important for the victims than usually mentioned by the rehabilitation team functional rehabilitation using all the modern facilities including surgical procedures has demonstrated the realistic possibility to have a better control of the lesion, to activate compensatory mechanisms and to develop efficient strategies for rehabilitative procedures and so to achieve today an acceptable final outcome in respect to the WHO-ICF definition.

In fact those muscles under the spinal cord lesion are still alive and even if they are not under brain control, they keep their contraction capacity, yet diminished, but clinically useably. Therefore Functional electrical stimulation (FES) and treadmill training [10] are the most efficient physical therapy to prevent muscle atrophy. FES was demonstrated to restore locomotion by transcutaneous [11] as well as by perineural and epimyseal stimulation using a neurprosthesis. The latter has been demonstrated recently in our SUAW project, stand up and walk [17].

Surgical procedures for plastic reconstructive surgery are still experimental in humans, as it was the SUAW project. The senior author GAB has performed and reported on his first patients in whom he was able to successfully restore voluntary muscle contractions, standing and walking after complete SCI by transferring the ulnar nerve to the hip to be connected with the distal moror branches of Nn gluteus maximus (for hip extension), gluteus medius (for abduction and stabilisation of the pelvis) and quadriceps femoris (for hip flexion in orthostasis and knee extension), based on extensive research in animals. The ulnar nerve was freed and cut in the canal of Gyon to be elevated from distal to its origin from the brachial plexus and afterwards guided down to the hip to be connect with sutures end to the motor branches which have been cut from the siatic nerve and the femoral nerve: the motor branches for the hypothenar and interossei (fascicles group 1), the adductor pollicis (no 2), and the two medial heads of the flexor digitorum profundi and the flexor carpi ulnaris (no 3) when the first two groups were connested with the Mm glutaeus maximus and gluteus medius, while the 3rd group was sutured with the femoral nerve to the m.quadriceps femoris. A second team performed contemporary the reconstructive surgery at the hand (Bunnel operation for adductor pollicis and R. Smith operation for interossei) which gave a functional satisfying motoric result to the donor hand. After 2 months the second procedure was performed on the other side, followed, after complete healing of the wounds, by meticulous physical education, first in bed and afterward in the gymnastic room and swimming pool. About 18 months proceeded before the muscles were innervated for muscle respons because of the wallerian degeneration. Initially the patient had to intend to move his hand in order to obtain contraction of the muscles of the hip, but thereafter the brain learned the new lessons and the patient was able to activate the muscles intentionally, however, by adding the voluntary command for the muscles of the hand, contraction increased. This was demonstrated with the electromyogram. Five patients altogether were operated with this technique. One patient became able to stand up from his wheelchair by himself and to walk for a short distance with help of a walker from one room to another and to reach the car and to drive. He feels again completely autonomous. The others however gave up their intensive physical training and specially tailored educational programme because they found it too time-consuming. We therefore believe that in all potential paraplegic candidates for all kinds of plastic-reconstructive surgery an extensive and repeted neurosychological preoperative assessment is absolutely mandatory.

If partial restoration of neurological function in complete SCI paraplegia cannot be observed within the first year we have to consider new techniques for re-engineering of the damaged spinal cord. E.g. *Brunelli's paradigme* where autologous nerve grafts are inserted into the ventro lateral cortical-spinal tract to be connected, by-passing the spinal cord lesion, with peripheral nerves for re-innervation of muscle groups responsible for locomotion. Despite the functional promising results in our patient after 27 month of reconstructive surgery.

Several uncertainties remain to be discussed in more detail to be clarified in the future. For instance, the fact that the muscles can be stimulated by the upper motor neuron (glutamatergic), whereas physiologically the motor endplates normally are in contact with the terminal of the cholinergic lower, the second motor neuron. We are searching now to know if the receptors changed from cholinergic to glutamatergic of if it is dependent on the upper first motor neuron which changed its neurotransmitter to acethylcholine instead of glutamate. A new research is in advanced progress to find out the genes codifying for motor endplates receptors after connection to the PNS.

Who might benefit from this kind of surgery? Which are the selection criteria?

When should this restorative operative procedure best be performed in paraplegic patients?

Our indications are similar to those we followed with the SUAW group [17]:

1. Victims with a complete lesion (group A) T8–T11 because of a total disconnection (disruption) of the spinal cord (MRI, TCMS, SSEPs) and, clinically no symptoms and signs of partial recovery over a period of no more than one year, with functional and radiological stable spinal column.
2. Patients should otherwise be healthy, psychological stable, fully independent in their wheelchair regarding their daily activities, and aware about their situation in respect to their expected permanent functional neurological deficit and life long burden of permanent handicap.
3. Patients should be willing to accept intensive physical training before and two years after the operation to be continued through out their life.
4. Patients should not have severe atrophy of the muscles of shoulder, back, upper limbs, and chest to guarantee sitting in an upright position and fixation of the body with both arms standing while standing up, standing and wolking with help of the walker.
5. Patient's body weight should be within the normal range plus maximal 15% kg to guarantee standing and walking with the walker or crutches.
6. Age limits are 18 to 40 years of age, the younger the better.

Regarding our SUAW project, however, most of the patients who wanted to be operated and were trained preoperatively with external FES suffered from SCI lesions older than five years. The two patients we have implanted the neuroprosthesis so far successfully suffered from complete cord lesion since 10 years and had severe muscle atrophy. The preoperative training of

3 months could not compensate the severe muscle deficits as otherwise expected. The earlier the better will give the best results, but we also may expect good functional restoration after years as long as the muscles can be stimulated and trained future. In reviewing the literature and our limited clinical results we are convinced that surgery will be the only chance to restore locomotion in SCI paraplegics for the next decade and a challenge for surgeons interested in interdisciplinary functional rehabilitation and re-engineering. An international multi centre study including 40 paraplegics is on the way to proore Brunelli's paradigme.

References

1. Bigamini A, Chir HN, Dhal D (eds) (1981) Posttraumatic peripheral nerve regeneration. Raven Press, pp 397–402
2. Bregman BS, Kunkel-Bagdan E, Schnell L, Dai HN, Gao D, Schwab M (1995) Recovery from spinal cord injury mediated by antibodies to neurite. Nature 378: 498–501
3. Brown AG (1981) Organization in the spinal cord. The anatomy and physiology of identified neurones. Springer, Berlin Heidelberg New York Tokyo
4. Brunelli G (1980) Nerve suture. Intern Surg 65(6): 499–501
5. Brunelli GA, Milanesi S (1988) Experimental repair of spinal cord lesions by grafting from CNS to PNS. J Rreconstr Microsurg 4: 245–250
6. Brunelli GA, Brunelli GR (1997) Upper limb to lower limb nerve transfer. Reconstr Microsurg 13: 263–265
7. Brunelli GA, Brunelli GR (1999) Restorationation walking in paraplegia by transferring the ulnar nerve to the hip: a report on the first patient. Reconstr Microsurg 13: 263–265
8. Carlstedt T, Grane P, Hallin RG *et al* (1999) Return of function after spinal cord implantation of avulsed spinal nerve roots. Lancet 346: 1323–1325
9. Cheng H, Cao Y, Olson L (1996) Spinal cord repair in adult paraplegic rats: partial restoration of hind limb function. Science 273: 510–513
10. Dietz V (1966) Querschnittslähmung. Kohlhammer, Stuttgart
11. Kobetic R, Triolo RJ, Moralais EB (1997) Muscle selection and walking performance of multichannel FES system for ambulation in paraplegia. IEEE Transactions on Rehabilitation Engineering 5: 23–29
12. Mautes AEM, Steudel W-J, Martin ES (2002) Actual aspects of treatment strategies in spinal cord injury. Eur J Trauma 3: 143–156
13. Massini M (2002) Scientific session on neurosurgical rehabilitation WFNS Committee. XXIII Congress of Brazilian Neurosurgical Society Fortaleza, Brazilia, September 2002
14. Privat *et al* (2002) Clinical research in spine trauma. In 3rd Deauville int conference on spinal cord report. IRME, Paris, France, pp 8–10
15. Richardson PM, MC Guiness UM, Aguanaio AJ (1980) Nature 284: 264
16. Tator ChH (96) Classification of spinal cord injury based on neurologiocal presentation. Neurotrauma. In: Narayan RK, Wilberger JE, Povlishock JT (eds) McGraw-Hill Comp Inc, USA, pp. 1059–073
17. von Wild K, Rabischong P, Brunelli G, Benichou M, Krishnan K (2001) Computer added locomotion by implanted electrical stimulation in paraplegic patients (SUAW). Acta Neurochir (Wien) [Suppl] 79: 99–104
18. Young W (1996) Spinal cord injury pathophysiology and therapy. Neurotrauma. In: Narayan RK, Wilberger JE, Povlishock JT (eds) McGraw-Hill Comp Inc, USA, pp 1075–1093

Correspondence: Prof. Klaus von Wild, Medical Faculty of the Westfälische Wilhelms University; Frauenburgstr. 32, D-48155 Münster, Germany. E-mail: kvw@neurosci.de

Neurosurgical treatment for movement disorders

Acta Neurochir (2003) [Suppl] 87: 115–120

Brain stimulation: history, current clinical application, and future prospects

A. Y. Mogilner[1] and **A. R. Rezai**[2]

[1] Department of Neurosurgery, New York Medical College, Valhalla, New York, U.S.A.
[2] Department of Neurosurgery, Cleveland Clinic Foundation, Cleveland, OH, U.S.A.

Summary

The dramatic effects of chronic brain stimulation in the treatment of movement disorders have spurred a renewed interest in this technique for treating a variety of other conditions. This technique has only recently begun to reach its vast clinical potential, due to a number of significant advances in basic and clinical neurosciences. Current image-guided navigation systems and intraoperative physiological mapping techniques offer more efficient, consistent, and precise targeting. Advances in neurophysiology have helped elucidate the pathophysiology of a number of disease states and thus provided for rational target selection for therapy. The latest generation of stimulation equipment allows for precise tailoring of stimulation parameters to maximize clinical benefit. These techniques are now being applied to a variety of other conditions including chronic pain, epilepsy, and psychiatric disorders.

Keywords: Brain stimulation; movement disorders; chronic pain.

Historical overview

In 1948, J. L. Pool of Columbia University placed a silver electrode, via open craniotomy, into the caudate nucleus of a woman with severe depression secondary to advanced Parkinson's disease, and connected it to an implanted induction coil [50]. Stimulation performed over a period of eight weeks improved her mood and her appetite. Chronic stimulation of a number of subcortical targets was used throughout the next decade to treat both psychiatric disorders and chronic pain [24]. Stereotactic neurosurgery, introduced by Spiegel and Wycis in 1947 [59], provided a less invasive and precise method of placing electrodes into the brain, and the vast majority of these procedures were performed stereotactically.

Interestingly, movement disorder surgery was one of the few areas where chronic stimulation was not considered a treatment option, due to the excellent clinical results obtained by subcortical lesioning procedures such as thalamotomy and pallidotomy. However, as stimulation was routinely used prior to creation of a permanent lesion in order to predict its clinical efficacy, the effects of stimulation on various movement disorders were well known. Spiegel and Wycis reported that low frequency stimulation of the globus pallidus could elicit or augment tremor [58], while Hassler noted that higher frequency stimulation could reduce or even arrest tremor [21, 23]. With the advent of levodopa in the 1960s, the use of stereotactic surgery for movement disorders dramatically declined. During the next two decades, the primary use of brain stimulation was for the treatment of refractory pain, with mixed results. With the introduction, by Benabid et al. [6] in 1987, of thalamic stimulation for tremor control, a renaissance in the use of brain stimulation was begun. The current deep brain stimulation (DBS) system is shown in Fig. 1. The electrodes are stereotactically implanted into the various targets of interest and connected to remotely controlled and programmable subcutaneously implanted pulse generators.

Movement disorders

Thalamic stimulation for tremor

Benabid's initial indication for chronic thalamic stimulation was in patients with bilateral refractory tremor, primarily essential tremor and parkinsonian tremor [6]. As bilateral thalamotomy is associated with increased morbidity including speech dysfunction [41], a contralateral non-destructive procedure following unilateral thalamotomy was envisioned as appropriate. Benabid and colleagues were the first to recognize the importance of high frequency stimulation (130 Hz or greater) for a consistent beneficial effect. At the

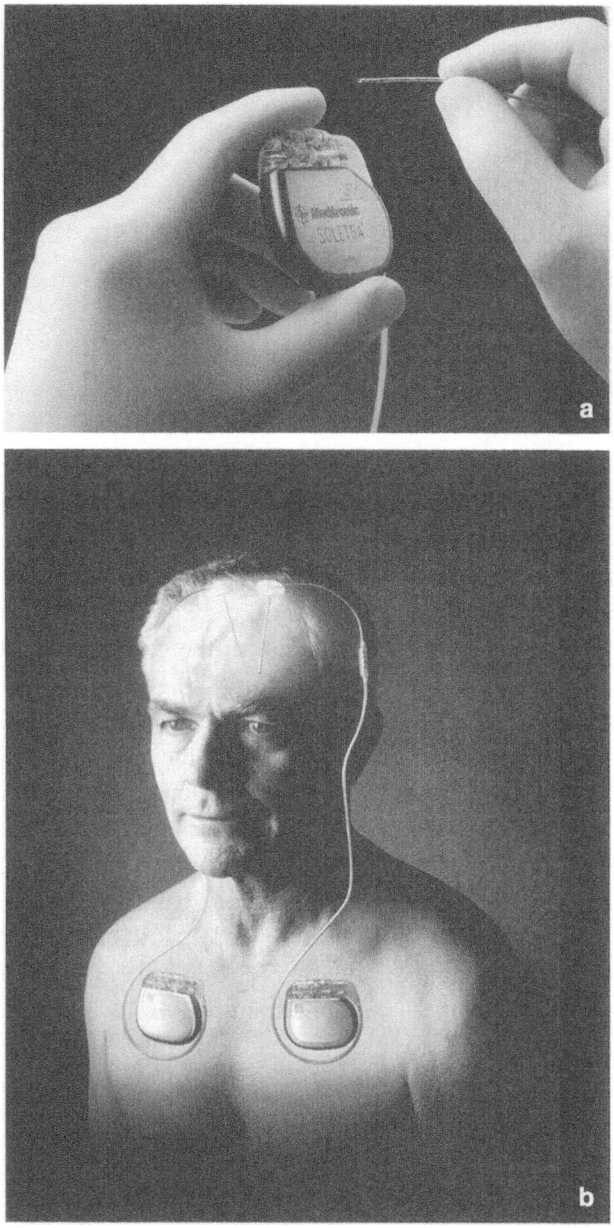

Fig. 1. (a) Photograph of the Soletra programmable pulse generator and the deep brain stimulating electrode. (b) Illustration of the DBS system showing the electrode, connecting wires, and pulse generators

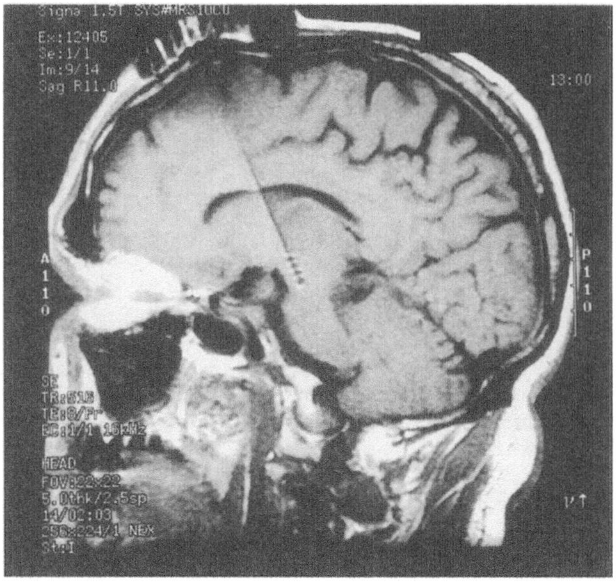

Fig. 2. Sagittal T1 weighted post-operative MRI scan showing DBS lead in the subthalamic nucleus

present time, Vim thalamic DBS has been established as a safe and effective modality for the treatment of Parkinsonian and essential tremor of the upper extremity, with equal efficacy to thalamotomy but with fewer complications. Approximately 80–90% of patients with Parkinson's disease and essential tremor obtain significant benefit from the procedure. Other parkinsonian symptoms, including rigidity, gait disturbance, bradykinesia, as well as drug-induced dyskinesias, however, are not significantly improved by thalamic stimulation [2, 5, 27, 39, 56].

Pallidal stimulation

As pallidotomy was known to improve multiple parkinsonian symptoms including rigidity, bradykinesia, gait dysfunction, tremor, as well as reduce the severity of levodopa-induced dyskinesias [33], the GPi became the next target for chronic stimulation. Pallidal stimulation has been shown to be a safe and effective means of treating refractory Parkinson's Disease [7, 18, 19, 71].

Subthalamic nucleus (STN) stimulation

Unlike the Vim thalamus and globus pallidus, the subthalamic nucleus (STN) was previously not considered for stereotactic lesioning procedures in Parkinson's disease. This was due to fear of causing hemiballism, a well-documented effect of STN lesions in previously normal individuals experiencing intracerebral hemorrhage in this area [34]. However, after primate models of Parkinson's disease demonstrated that STN lesions could alleviate many of the symptoms of Parkinson's Disease, Benabid and colleagues [38] introduced STN stimulation in 1993, and described the first case series in 1995. Since that time, the

STN has become the most common target for chronic electrical stimulation (Fig. 2). STN DBS has been shown to be effective in treating the cardinal manifestations of the disease including rigidity, bradykinesia/akinesia, gait disturbance, as well as tremor, with significant improvements in overall motor function using the standardized UPDRS rating scale. Furthermore, levodopa-induced dyskinesias are significantly reduced post-operatively, both due to a significant reduction in levodopa intake and via a direct anti-dyskinetic effect of stimulation [28, 29, 36–38, 49].

Other movement disorders

Thalamic and pallidal stimulation have been shown to be efficacious for various forms of dystonia, with more beneficial effects noted for primary and generalized dystonias as opposed to secondary and focal syndromes [26, 30, 32, 42, 57, 61, 62]. Vandewalle et al. reported relief of motor tics in a patient with Tourette's syndrome with bilateral medial thalamic DBS [67].

Chronic pain

Initial work by Pool, Heath and others involved stimulation of a variety of subcortical targets for pain control and were viewed more as a psychosurgical intervention aimed at treating the so-called "affective" state of the individual. Subsequently, two diencephalic targets were introduced – sensory thalamus (VC, VP) and periacqueductal/periventricular gray (PAG/PVG). Sensory thalamic stimulation, first reported by Mazars and colleagues [43–45], produces paresthesias in the area of pain, similar to that obtained with stimulation of the dorsal columns of the spinal cord. Stimulation of the PAG/PVG, first reported in 1969, provided yet another proposed pathway of analgesia, via a presumed opioid-related mechanism [52–54]. PAG/PVG stimulation is considered a better target for nociceptive pain, while sensory thalamus has been targeted for deafferentation or neuropathic pain conditions [35]. There are few long-term outcome studies of DBS for pain. Kumar *et al.* reported pain relief in approximately 60% of patients with nociceptive pain and 50% of those with neuropathic pain with a mean 78 month follow-up [31], while Levy *et al.* reported long-term pain relief in approximately 30% of patients [35].

Epidural motor cortex stimulation (MCS), introduced by Tsubokawa and colleagues in the early 1990s, is being increasingly used to treat refractory pain syndromes. While attempting sensory cortical stimulation for pain control, Tsubokawa and colleagues found, perhaps fortuitously, that stimulation of the precentral motor cortex, rather than the somatosensory postcentral cortex, resulted in effective pain relief [63, 64]. Stimulation parameters were adjusted to be below the threshold for a motor response. The analgesic effects had a "halo" effect, lasting at times for hours after stimulation was discontinued.

An increasing number of groups are now reporting promising results with this technique (For a recent multi-article review, see Neurosurgical Focus, 11:3.). The largest series comprises 32 patients spanning a 4-year period, reported by Nguyen and colleagues [46]. 77% of patients with central pain and 75% of patients with neuropathic facial pain had experienced substantial pain relief. Unlike paresthesia-producing (i.e., sensory thalamic stimulation), MCS does not usually induce paresthesias, making this technique ideal for double-blinded controlled studies.

The mechanism of action of MCS is as yet unclear. Drouot and colleagues recently reported that patients with good response to MCS display either an absence of alteration of non-nociceptive (thermal) sensory modalities within the painful area, or abnormal sensory thresholds that improve with MCS [16]. These results suggest that MCS effects pain relief by modulating the activity of sensory cortical areas. While some have postulated that direct projections from motor cortex to sensory cortex can provide for such sensory modulation, others suggest that descending corticothalamic projections play an important role in the generation of analgesia [48, 63, 64].

Epilepsy

Initial attempts at seizure control using neurostimulation utilized stimulation of the cerebellar cortex. Cooper and colleagues [9–13] were the first to employ cerebellar stimulation for epilepsy. While he reported significant seizure reduction in 56% of patients, other controlled clinical trials failed to demonstrate any benefit [66, 72] and the technique was largely abandoned.

At the present time, the most frequently used stimulation target for seizure control is the extracranial vagus nerve [20]. While vagal stimulation has been shown to effect a moderate reduction in seizure frequency, studies of intracranial targets continue in an attempt to obtain improved clinical outcomes [40].

Three such targets currently under investigation include the anterior nucleus of the thalamus, the centromedian nucleus of the thalamus, and the subthalamic nucleus.

Anterior nucleus stimulation was first reported by Cooper et al. [14, 15]. Although exact details of the stimulation were not provided, they reported that in five of 6 patients with intractable seizures, stimulation reduced seizures by more than 60% in 5 of 6 cases, with 30% average decrease in medication requirements. A multicenter study is currently underway re-evaluating anterior nucleus stimulation. Velasco *et al.* were the first to employ bilateral centromedian (CM) nucleus stimulation for the treatment of epilepsy [69], and their long-term results in thirteen patients have been reported [68]. Stimulation parameters were 60 Hz, 4–6 V, alternating right and left sides. With a mean follow-up of 41 months, they noted a significant decrease (defined as >80% seizure reduction) in the incidence of generalized tonic-clonic seizures and atypical absence seizures, but no change in the incidence of complex partial seizures.

The STN has recently become a target for epilepsy control. The theoretical basis for controlling seizures with STN stimulation derives from its projections to the substantia nigra pars reticulata (SNr), which has been shown to exert significant inhibitory control over seizure propagation in animal models [17, 25]. On that basis, high-frequency STN stimulation was demonstrated to suppress seizure activity in the rat [70]. Benabid and colleagues recently reported significant improvement in a 5-year-old female following STN stimulation [4]. Preliminary results from both Grenoble and the Cleveland Clinic are promising, but stress the need for further studies [3, 40].

Other emerging areas

Psychiatric disorders

Since 1948, when Pool treated psychiatric disorders with brain stimulation, there have been few advances in the use of brain stimulation to treat psychiatric conditions. While lesioning procedures including cingulotomy, capsulotomy, subcaudate tractotomy and limbic leukotomy have all been employed for the treatment of refractory obsessive-compulsive disorder (OCD) and, to a lesser degree, major affective disorder, chronic electrical stimulation of these lesioning targets has only recently been reported. Nuttin *et al.* placed quadripolar electrodes bilaterally in the anterior limbs of the internal capsules of four patients suffering from severe OCD. Beneficial effects were seen in three patients [47]. A multi-center study using stimulation of the anterior limb of the internal capsule is now underway. A recent report of transient depression induced by high-frequency stimulation of the substantia nigra pars reticulata [1], using the lower poles of an electrode placed for subthalamic nucleus stimulation, raise the intriguing possibility that STN and/or SNr stimulation may be a putative target for future intervention.

Eating disorders

Over twenty-five years ago, Quaade *et al.* performed low-frequency stimulation of the lateral hypothalamic nucleus in severely obese patients as a prelude to electrocoagulation, without significant weight reduction [51]. As Benabid and colleagues have reported preliminary work with both lateral and ventromedial hypothalamic stimulation in an animal model [3], others may follow this lead in investigating this potential application of brain stimulation.

Brain injury states

A number of targets have been stimulated in vegetative patients in an attempt to increase their level of consciousness. The midbrain reticular formation, intralaminar thalamus, other thalamic nuclei, and the globus pallidus all have been stimulated with mixed results [8, 22, 60, 65]. A recent proposal to revisit intralaminar stimulation in a less severely disabled group of patients, described as minimally conscious [55], appears promising and a multi-center study of this technique will soon be underway.

Conclusions

It is important to emphasize that, despite the dramatic clinical results obtainable from brain stimulation, the technique is still in its infancy. Advances in our understanding of the pathophysiology of the disorders, improvements in brain stimulation device technology, and the development of safer and less invasive surgical techniques will help this technique reach its clinical and therapeutic potential.

References

1. Bejjani BP, Damier P, Arnulf I *et al* (1999) Transient acute depression induced by high-frequency deep-brain stimulation [see comments]. N Engl J Med 340: 1476–1480
2. Benabid AL, Benazzouz A, Hoffmann D *et al* (1998) Long-term electrical inhibition of deep brain targets in movement disorders. Mov Disord 13: 119–125
3. Benabid AL, Koudsié A, Pollak P *et al* (2000) Future prospects of brain stimulation. Neurol Res 22: 237–246
4. Benabid AL, Minotti L, Koudsie A *et al* (2002) Antiepileptic effect of high-frequency stimulation of the subthalamic nucleus (corpus luysi) in a case of medically intractable epilepsy caused by focal dysplasia: a 30-month follow-up: technical case report. Neurosurgery 50: 1385–1391; discussion 1391–1382
5. Benabid AL, Pollak P, Gao D *et al* (1996) Chronic electrical stimulation of the ventralis intermedius nucleus of the thalamus as a treatment of movement disorders. J Neurosurg 84: 203–214
6. Benabid AL, Pollak P, Louveau A *et al* (1987) Combined (thalamotomy and stimulation) stereotactic surgery of the VIM thalamic nucleus for bilateral Parkinson disease. Appl Neurophysiol 50: 344–346
7. Burchiel KJ, Anderson VC, Favre J *et al* (1999) Comparison of pallidal and subthalamic nucleus deep brain stimulation for advanced Parkinson's disease: Results of a randomized, blinded pilot study. Neurosurgery 45: 1375–1384
8. Cohadon F, Richer E (1993) Deep cerebral stimulation in patients with post-traumatic vegetative state. 25 cases. Neurochirurgie 39: 281–292
9. Cooper IS (1978) Cerebellar stimulation in man. Raven Press, New York
10. Cooper IS (1973) Effect of chronic stimulation of anterior cerebellum on neurological disease. Lancet 1: 206
11. Cooper IS (1973) Effect of stimulation of posterior cerebellum on neurological disease. Lancet 1: 1321
12. Cooper IS, Amin I, Gilman S (1978) The effect of chronic cerebellar stimulation upon epilepsy in man. Trans Am Neurol Assoc 98: 192–196
13. Cooper IS, Amin I, Riklan M *et al* (1976) Chronic cerebellar stimulation in epilepsy. Arch Neurol 33: 559–570
14. Cooper IS, Upton AR (1985) Therapeutic implications of modulation of metabolism and functional activity of cerebral cortex by chronic stimulation of cerebellum and thalamus. Biol Psychiatry 20: 811–813
15. Cooper IS, Upton AR, Amin I (1980) Reversibility of chronic neurologic deficits. Some effects of electrical stimulation of the thalamus and internal capsule in man. Appl Neurophysiol 43: 244–258
16. Drouot X, Nguyen JP, Peschanski M *et al* (2002) The antalgic efficacy of chronic motor cortex stimulation is related to sensory changes in the painful zone. Brain 125: 1660–1664
17. Gale K, Iadarola MJ (1980) Seizure protection and increased nerve-terminal GABA: delayed effects of GABA transaminase inhibition. Science 208: 288–291
18. Galvez-Jimenez N, Lozano A, Tasker R *et al* (1998) Pallidal stimulation in Parkinson's disease patients with a prior unilateral pallidotomy. Can J Neurol Sci 25: 300–305
19. Ghika J, Villemure JG, Fankhauser H *et al* (1998) Efficiency and safety of bilateral contemporaneous pallidal stimulation (deep brain stimulation) in levodopa-responsive patients with Parkinson's disease with severe motor fluctuations: a 2-year follow-up review. J Neurosurg 89: 713–718
20. Group TVNS (1995) A randomized controlled trial of chronic vagus nerve stimulation for treatment of medically intractable seizures. The Vagus Nerve Stimulation Study Group. Neurology 45: 224–230
21. Hassler R (1966) Thalamic regulation of muscle tone and the speed of movements. In: Purpura DP, Yahr MD (eds) The thalamus. Columbia University Press, New York
22. Hassler R, Ore GD, Bricolo A *et al* (1969) EEG and clinical arousal induced by bilateral long-term stimulation of pallidal systems in traumatic vigil coma. Electroencephalogr Clin Neurophysiol 27: 689–690
23. Hassler R, Riechert TFM (1960) Physiological observations in stereotaxic operations in extrapyramidal motor disturbances. Brain 83: 337–350
24. Heath R, Mickle W (1960) Evaluation of 7 years' experience with depth electrode studies in human patients. In: Ramey E, O'Doherty D (eds) Electrical studies in unanesthetized brain. Harper & Brothers, New York
25. Iadarola MJ, Gale K (1982) Substantia nigra: site of anticonvulsant activity mediated by gamma-aminobutyric acid. Science 218: 1237–1240
26. Islekel S, Zileli M, Zileli B (1999) Unilateral pallidal stimulation in cervical dystonia. Stereotact Funct Neurosurg 72: 248–252
27. Koller W, Pahwa R, Busenbark K *et al* (1997) High-frequency unilateral thalamic stimulation in the treatment of essential and parkinsonian tremor. Ann Neurol 42: 292–299
28. Krack P, Limousin P, Benabid AL *et al* (1997) Chronic stimulation of subthalamic nucleus improves levodopa-induced dyskinesias in Parkinson's disease [letter]. Lancet 350: 1676
29. Krack P, Pollak P, Limousin P *et al* (1997) Stimulation of subthalamic nucleus alleviates tremor in Parkinson's disease [letter]. Lancet 350: 1675
30. Krauss JK, Pohle T, Weber S *et al* (1999) Bilateral stimulation of globus pallidus internus for treatment of cervical dystonia [letter]. Lancet 354: 837–838
31. Kumar K, Toth C, Nath RK (1997) Deep brain stimulation for intractable pain: a 15-year experience. Neurosurgery 40: 736–746; discussion 746–737
32. Kumar R, Dagher A, Hutchison WD *et al* (1999) Globus pallidus deep brain stimulation for generalized dystonia: clinical and PET investigation. Neurology 53: 871–874
33. Laitinen L, AT B, Hariz M (1992) Leksell's posteroventral pallidotomy in the treatment of Parkinson's disease. J Neurosurg 76: 53–61
34. Lee MS, Marsden CD (1994) Movement disorders following lesions of the thalamus or subthalamic region. Mov Disord 9: 493–507
35. Levy RM, Lamb S, Adams JE (1987) Treatment of chronic pain by deep brain stimulation: long term follow-up and review of the literature. Neurosurgery 21: 885–893
36. Limousin P, Krack P, Pollak P *et al* (1998) Electrical stimulation of the subthalamic nucleus in advanced Parkinson's disease. N Engl J Med 339: 1105–1111
37. Limousin P, Pollak P, Benazzouz A *et al* (1995) Bilateral subthalamic nucleus stimulation for severe Parkinson's disease. Mov Disord 10: 672–674
38. Limousin P, Pollak P, Benazzouz A *et al* (1995) Effect of parkinsonian signs and symptoms of bilateral subthalamic nucleus stimulation. Lancet 345: 91–95
39. Limousin P, Speelman JD, Gielen F *et al* (1999) Multicentre European study of thalamic stimulation in parkinsonian and essential tremor. J Neurol Neurosurg Psychiatry 66: 289–296
40. Loddenkemper T, Pan A, Neme S *et al* (2001) Deep brain stimulation in epilepsy. J Clin Neurophysiol 18: 514–532
41. Louw DF, Burchiel KJ (1998) Ablative therapy for movement disorders. Complications in the treatment of movement disorders. Neurosurg Clin N Am 9: 367–373

42. Lozano AM, Kumar R, Gross RE *et al* (1997) Globus pallidus internus pallidotomy for generalized dystonia [see comments]. Mov Disord 12: 865–870
43. Mazars GL (1975) Intermittent stimulation of nucleus ventralis posterolateralis for intractable pain. Surg Neurol 4: 93–95
44. Mazars GL, Merienne L, Ciolocca C (1973) Stimulations thalamiques intermittents antalgiques. Note Préeiminaire. Revue Neurol 128: 273–279
45. Mazars GL, Merienne L, Ciolocca C (1974) Treatment of certain types of pain by implantable thalamic stimulators. Neurochirurgie 29: 117–124
46. Nguyen JP, Lefaucheur JP, Decq P *et al* (1999) Chronic motor cortex stimulation in the treatment of central and neuropathic pain. Correlations between clinical, electrophysiological and anatomical data. Pain 82: 245–251
47. Nuttin B, Cosyns P, Demeulemeester H *et al* (1999) Electrical stimulation in anterior limbs of internal capsules in patients with obsessive-compulsive disorder [letter]. Lancet 354: 1526
48. Peyron R, Garcia-Larrea L, Deiber MP *et al* (1995) Electrical stimulation of precentral cortical area in the treatment of central pain: electrophysiological and PET study. Pain 62: 275–286
49. Pollak P, Benabid AL, Limousin P *et al* (1996) Subthalamic nucleus stimulation alleviates akinesia and rigidity in parkinsonian patients. Adv Neurol 69: 591–594
50. Pool JL (1954) Psychosurgery in older people. J Am Geriat Soc 2: 456–465
51. Quaade F, Vaernet K, Larsson S (1974) Stereotaxic stimulation and electrocoagulation of the lateral hypothalamus in obese humans. Acta Neurochir (Wien) 30: 111–117
52. Reynolds D (1969) Surgery in the rat during electrical analgesia induced by focal brain stimulation. Science 164
53. Richardson DE, Akil H (1977) Pain reduction by electrical brain stimulation in man. Part 1: Acute administration in periaqueductal and periventricular sites. J Neurosurg 47: 178–183
54. Richardson DE, Akil H (1977) Pain reduction by electrical brain stimulation in man. Part 2: Chronic self-administration in the periventricular gray matter. J Neurosurg 47: 184–194
55. Schiff ND, Rezai AR, Plum F (2000) A neuromodulation strategy for rational therapy of complex brain injury states. Neurol Res 22: 267–272
56. Schuurman PR, Bosch DA, Bossuyt PM *et al* (2000) A comparison of continuous thalamic stimulation and thalamotomy for suppression of severe tremor [see comments]. N Engl J Med 342: 461–468
57. Sellal F, Hirsch E, Barth P *et al* (1993) A case of symptomatic hemidystonia improved by ventroposterolateral thalamic electrostimulation. Mov Disord 8: 515–518
58. Spiegel EA, Wycis HT, Baird HW *et al* (1960) Physiopathologic observations on the basal ganglia. In: Ramey E, O'Doherty DS (eds) Electrical studies on the unanesthetized brain. Harper & Brothers, New York
59. Spiegel EA, Wycis HT, Marks M *et al* (1947) Stereotaxic apparatus for operations on the human brain. Science 106: 349–350
60. Sturm V, Kuhner A, Schmitt HP *et al* (1979) Chronic electrical stimulation of the thalamic unspecific activating system in a patient with coma due to midbrain and upper brain stem infarction. Acta Neurochir (Wien) 47: 235–244
61. Thompson TP, Kondziolka D, Albright AL (2000) Thalamic stimulation for choreiform movement disorders in children. Report of two cases. J Neurosurg 92: 718–721
62. Tronnier VM, Fogel W (2000) Pallidal stimulation for generalized dystonia. Report of three cases. J Neurosurg 92: 453–456
63. Tsubokawa T, Katayama Y, Yamamoto T *et al* (1991) Chronic motor cortex stimulation for the treatment of central pain. Acta Neurochir Suppl 52: 137–139
64. Tsubokawa T, Katayama Y, Yamamoto T *et al* (1993) Chronic motor cortex stimulation in patients with thalamic pain. J Neurosurg 78: 393–401
65. Tsubokawa T, Yamamoto T, Katayama Y *et al* (1990) Deep-brain stimulation in a persistent vegetative state: follow-up results and criteria for selection of candidates. Brain Inj 4: 315–327
66. Van Buren JM, Wood JH, Oakley J *et al* (1978) Preliminary evaluation of cerebellar stimulation by double-blind stimulation and biological criteria in the treatment of epilepsy. J Neurosurg 48: 407–416
67. Vandewalle V, van der Linden C, Groenewegen HJ *et al* (1999) Stereotactic treatment of Gilles de la Tourette syndrome by high frequency stimulation of thalamus [letter]. Lancet 353: 724
68. Velasco F, Velasco M, Jimenez F *et al* (2000) Predictors in the treatment of difficult-to-control seizures by electrical stimulation of the centromedian thalamic nucleus [In process citation]. Neurosurgery 47: 295–304; discussion 304–295
69. Velasco F, Velasco M, Ogarrio C *et al* (1987) Electrical stimulation of the centromedian thalamic nucleus in the treatment of convulsive seizures: a preliminary report. Epilepsia 28: 421–430
70. Vercueil L, Benazzouz A, Deransart C *et al* (1998) High-frequency stimulation of the subthalamic nucleus suppresses absence seizures in the rat: comparison with neurotoxic lesions. Epilepsy Res 31: 39–46
71. Volkmann J, Sturm V, Weiss P *et al* (1998) Bilateral high-frequency stimulation of the internal globus pallidus in advanced Parkinson's disease. Ann Neurol 44: 953–961
72. Wright G, McLellan D, Brice A (1984) A double-blind trial of chronic cerebellar stimulation in twelve patients with epilepsy. J Neurol Neurosurg Psychiatry 47: 769–774

Correspondence: Ali R. Rezai, M.D., Department of Neurosurgery, The Cleveland Clinic Foundation, 9500 Euclid Avenue, S80, Cleveland, OH 44195, U.S.A. E-mail: rezaia@ccf.org

Acta Neurochir (2003) [Suppl] 87: 121–123

Deep brain and motor cortex stimulation for post-stroke movement disorders and post-stroke pain

Y. Katayama, T. Yamamoto, K. Kobayashi, H. Oshima, and **C. Fukaya**

Department of Neurological Surgery and Division of Applied System Neuroscience, Nihon University School of Medicine, Tokyo, Japan

Summary

Our experience of deep brain stimulation (DBS) and motor cortex stimulation (MCS) in patients with post-stroke movement disorders and post-stroke pain is reviewed. DBS of the thalamic nuclei ventralis oralis posterior et intermedius proved to be useful in more than 70% of patients with post-stroke involuntary movements (hemiballismus, hemichoreo-athetosis, distal resting and/or action tremor, and proximal postural tremor). The effect of DBS of the thalamic nucleus ventralis caudalis or internal capsule on post-stroke pain was usually disappointing. Excellent pain control can be achieved by MCS in approximately 50% of patients with post-stroke pain. In the course of clinical trials on MCS for the control of post-stroke pain, it was found that co-existent post-stroke involuntary movements (hemichoreo-athetosis and resting tremor) could also be controlled by MCS. Post-stroke involuntary movements, especially those in thalamic syndrome, are sometimes associated with post-stroke pain. In such disorders, involuntary movements are attenuated, but the pain in the same patients is often exacerbated by DBS of the thalamic nuclei ventralis oralis posterior et intermedius. MCS could be the therapy of choice under such circumstances. Subjective improvement of voluntary motor performance, which had been impaired in association with mild or moderate hemiparesis, was reported during MCS by approximately 20% of patients with post-stroke pain. Such an effect on voluntary motor performance appears to be caused by an inhibition of their rigidity. The reversibility of DBS and MCS makes them an important option for the control of post-stroke movement disorders and post-stroke pain.

Keywords: Brain stimulation; hemiballismus; hemichoreo-athetosis; involuntary movement; motor cortex; stroke; thalamus; tremor.

Introduction

Post-stroke movement disorders are difficult to control medically if they persist. Post-stroke pain is one of pain which is most difficult to control once it occurs. Such sequelae of stroke greatly limit the patient's activity of daily living. Thalamotomy and pallidotomy could serve to attenuate post-stroke movement disorders. Thalamotomy has also been reported to reduce post-stroke pain. It is not always certain, however, whether such ablative procedures provide sufficient benefit or not to justify creating irreversible lesions. Ablative procedures always carry a risk of complications associated with the creation of additional lesions within the brain which is already damaged by stroke. For this reason, thalamotomy and pallidotomy have not become popular for use in the control of post-stroke movement disorders and post-stroke pain.

During the last decade, it has become clear that deep brain stimulation (DBS) can provide a dramatic improvement in the symptoms of various movement disorders [2–4, 11, 12], including post-stroke involuntary movements [1, 5, 14]. DBS has also been used for the control of post-stroke pain [13]. Motor cortex stimulation (MCS) is another approach for controlling post-stroke movement disorders and pain [7–10, 13]. In contrast to ablative procedures, DBS and MCS do not carry a risk of complications associated with the creation of additional lesions. We have therefore been employing DBS and MCS to control post-stroke involuntary movements and post-stroke pain. In this report, our experience of DBS and MCS in patients with movement disorders and pain after stroke is reviewed [7–10, 13, 14].

Deep brain stimulation

Attempts to control post-stroke involuntary movements with DBS have been made by several investigators. Cooper *et al.* [5] reported that clinically useful improvement was achieved after DBS in 6 of 9 cases with unspecified involuntary movements caused by stroke. Andy [1] also described a case in which thalamic syndrome with choreiform movements following stroke was successfully treated by DBS. We have reported 2 cases of hemiballismus caused by stroke in which DBS of the thalamic nucleus ventralis intermedius provided excellent control [14]. We subsequently confirmed with a larger number of patients that various types of post-stroke movement disorders, such as hemichoreo-athetosis, distal resting and/or action tremor, and proximal postural tremor, can be controlled by DBS of the thalamic nuclei ventralis oralis posterior et intermedius [7]. DBS was found to provide excellent control of these involuntary movements in more than 70% of patients [7]. No stimulation-related side effects were induced at the stimulation intensities required to achieve these results. The effects continued to be seen for the entire follow-up period ranging 1–10 years.

DBS of the thalamic nuclei ventralis oralis posterior et intermedius was performed with a quadripolar electrode (1.5 mm in length, each point separated by 1.5 mm; Medtronic Inc., USA) which was placed in such a way that the most distal contact was situated on the ventral portion of the thalamic nucleus ventralis intermedius. Stimulation was applied bipolarly with various combinations of electrodes and parameters to achieve the best effects in each individual patient. The stimulation parameters and contact points used for DBS were modified at each follow-up visit to our clinic, based on the results of neurological examinations as well as the patient's report concerning his or her activity of daily living.

DBS of the thalamic nucleus ventralis caudalis or internal capsule can sometimes cause pain reduction to some extent in patients with post-stroke pain. The effect of DBS on the post-stroke pain was, however, usually disappointing [13]. The pain in many patients even became exacerbated following DBS of the thalamic nucleus ventralis caudalis. It is interesting to note that such a response to stimulation of the thalamic nucleus ventralis caudalis is rarely encountered in non-pain patients.

Motor cortex stimulation

We previously noted that excellent pain control can be achieved in approximately 50% of post-stroke pain patients undergoing MCS [13]. In the remaining patients, the stimulation system was not internalized because no pain control was reported during the test stimulation period, or pain control was obtained initially but the effect faded away over several months after internalization of the stimulation system. The area providing strong pain inhibition appeared to correspond to the motor cortex, since such an area caused muscle contraction if stimulated at a higher intensity. We confirmed subsequently with a larger number of patients that post-stroke pain could be controlled by MCS more frequently in patients whose motor weakness is not too severe [8].

MCS was performed with an electrode array consisting of 4 plate electrodes (5 mm in diameter, each separated by 5 mm; Medtronic Inc., USA), which was put in place through a small craniotomy. In general, the location of the motor cortex was determined from muscle contraction in response to bipolar stimulation with electrical pulses of low frequency. Stimulation was applied bipolarly with various combinations of electrodes and parameters to achieve the best effects in each individual patient. The intensity was restricted to a level slightly lower than the threshold for muscle contraction.

In the course of clinical trials on MCS for the control of post-stroke pain, we have encountered patients who were delighted by an unexpected improvement in their co-existent post-stroke involuntary movements during MCS [8, 10, 13]. It was found that hemichoreo-athetosis and resting tremor could be controlled by MCS [7, 9]. This observation suggests that MCS may be useful for controlling post-stroke involuntary movements similarly to DBS. Woolsey *et al.* [15] reported that, in severe Parkinson's disease, marked inhibition of the tremor and rigidity could be induced by MCS. The thalamic nucleus ventralis intermedius and motor cortex are the last relays of extrapyramidal neuronal loops to the corticospinal tract. It has long been recognized that post-stroke involuntary movements can be abolished when the corticospinal tract is interrupted. The effects of DBS and MCS on post-stroke involuntary movements may be related to a desynchronization of abnormal activities which are transmitted to the corticospinal tract.

Post-stroke involuntary movements, especially those in thalamic syndrome, are sometimes associated with post-stroke pain. DBS of the thalamic nuclei ventralis oralis posterior et intermedius can produce opposite effects in such disorders; involuntary movements are attenuated, but the pain in the same patients becomes exacerbated. This represents the problem which may be encountered in patients who suffer from post-stroke movement disorders and post-stroke pain at the same time. MCS could be the therapy of choice under such circumstances, since post-stroke pain may also be controlled by MCS [9].

During MCS for the control of post-stroke pain, some patients experience subjective improvement of their voluntary motor performance, which had been impaired in association with mild or moderate hemiparesis [7, 13]. We found subsequently with a larger number of patients that this phenomenon was reported by approximately 20% of patients during MCS [8, 9]. The effect was independent of pain control. In one patient who reported no pain control with MCS, the stimulation system was, nevertheless, internalized because he was so pleased by the marked improvement that occurred in his voluntary motor performance [7, 9]. Fields and Adams [6] found that DBS of the posterior limb of the internal capsule attenuated pain as well as motor deficits caused by cortical injury. Such an effect of MCS and DBS of the internal capsule on voluntary motor performance appears to be caused by an inhibition of their rigidity, since no increase in muscle strength was objectively detectable.

Conclusion

DBS and MCS are useful in many patients to control post-stroke movement disorders and post-stroke pain. The reversibility of the procedures makes them an important option for patients where ablative procedures should be minimized since their brain is already damaged by stroke.

Acknowledgment

This work was supported by grants from the Ministry of Education, Science and Culture of Japan (B09470302 and A12307029).

References

1. Andy OJ (1983) Thalamic stimulation for control of movement disorders. Appl Neurophysiol 46: 107–111
2. Benabid AL, Pollack P, Louveau A, Henry S, Rougemont JD (1987) Combined (thalamotomy and stimulation) stereotactic surgery of the VIM thalamic nucleus for bilateral parkinson disease. Appl Neurophysiol 50: 344–346
3. Blond S, Caparros-Lefebvre D, Parker F, Assaker R, Petit H, Guieu J-D, Christiaens J-L (1992) Control of tremor and involuntary movement disorders by chronic stereotactic stimulation of the ventral intermediate thalamic nucleus. J Neurosurg 77: 62–68
4. Brice J, McLellan L (1980) Suppression of intention tremor by contingent deep-brain stimulation. Lancet 1: 1221–1222
5. Cooper IS, Upton ARM, Amin I (1982) Chronic cerebellar stimulation (CCS) and deep brain stimulation (DBS) in involuntary movement disorders. Appl Neurophysiol 45: 209–217
6. Fields H, Adams JE (1974) Pain after cortical injury relieved by electrical stimulation of the internal capsule. Brain 97: 169–178
7. Katayama Y, Fukaya C, Yamamoto T (1997) Control of post-stroke involuntary and voluntary movement disorders with deep brain or epidural cortical stimulation. Stereotact Funct Neurosurg 69: 73–79
8. Katayama Y, Fukaya C, Yamamoto T (1998) Post-stroke pain control by chronic motor cortex stimulation: Neurological characteristics predicting favorable response. J Neurosurg 89: 585–591
9. Katayama Y, Oshima H, Fukaya C, Kawamata T, Yamamoto T (2001) Control of post-stroke movement disorders using chronic motor cortex stimulation. Acta Neurochir (Wien) [Suppl] 79: 89–92
10. Katayama Y, Tsubokawa T, Yamamoto T (1994) Chronic motor cortex stimulation for central deafferentation pain: Experience with bulbar pain secondary to Wallenberg syndrome. Stereotact Funct Neurosurg 62: 295–299
11. N'Guyen JP, Degos JD (1993) Thalamic stimulation and proximal tremor: A specific target in the Vim nucleus. Arch Neurol 50: 498–500
12. Siegfried J, Lippitz B (1994) Chronic electrical stimulation of the VL-VPL complex and of the pallidum in the treatment of movement disorders: Personal experience since 1982. Sterotact Funct Neurosurg 62: 71–75
13. Tsubokawa T, Katayama Y, Yamamoto T, Hirayama T, Koyama S (1993) Chronic motor cortex stimulation in patients with thalamic pain. J Neurosurg 78: 393–401
14. Tsubokawa T, Katayama Y, Yamamoto T (1995) Control of persistent hemiballismus by chronic thalamic stimulation. J Neurosurg 82: 501–505
15. Woolsey CN, Erickson T, Gilson WE (1979) Localization in somatic sensory and motor areas of human cerebral cortex as determined by direct recording of evoked potentials and electrical stimulation. J Neurosurg 51: 476–506

Correspondence: Yoichi Katayama, M.D., Ph.D., Department of Neurological Surgery, Nihon University School of Medicine, Itabashi-ku, Tokyo 173-8610, Japan. E-mail: ykatayam@med.nihon-u.ac.jp

Acta Neurochir (2003) [Suppl] 87: 125–128

Chronic simulation of the globus pallidus internus for control of primary generalized dystonia

Y. Katayama, C. Fukaya, K. Kobayashi, H. Oshima, and **T. Yamamoto**

Department of Neurological Surgery and Division of Applied System Neuroscience, Nihon University School of Medicine, Tokyo, Japan

Summary

Our experience of deep brain stimulation of the globus pallidus internus (GPi-DBS) for dystonia is summarized. A total of 5 patients with primary generalized dystonia underwent GPi-DBS. There were 3 males and 2 females. The age at onset of dystonia ranged from 8 to 45 years and the age at surgery for GPi-DBS ranged from 17 to 59 years. Two of the patients had been treated previously by bilateral thalamotomy or unilateral pallidotomy at other clinics and then developed new symptoms or recurrence. All were stimulated bilaterally. No surgical complications were encountered. The symptoms of dystonia were scored by the Burke-Fahn-Marsden dystonia rating scale (BFMDRS). The scores ranged from 18 to 62 before surgery. An improvement in the symptoms of dystonia was observed soon after the initiation of GPi-DBS, and additional progressive improvement was noted during a period of months or even years after surgery. The score at 6 months after surgery reached a level ranging from 4 to 23. The improvement in score ranged from −51% to −92%. GPi-DBS produced a marked effect even in patients who had previously undergone thalamotomy or pallidotomy. At 6 months after surgery, all patients were receiving bipolar stimulation with a wide interpolar distance, using contact 0 or 1 as the cathode and contact 2 or 3 as the anode. Stimulation was being performed at an intensity of around 2.0 V with a pulse width of 0.21 ms at a high frequency ranging from 120 to 140 Hz. GPi-DBS represents an important therapeutic option in many patients with primary generalized dystonia.

Keywords: Brain stimulation; dystonia; globus pallidus; involuntary movement; movement disorders.

Introduction

Dystonia is a movement disorder characterized by sustained muscle contractions leading to twisting movements and abnormal postures. The mechanism of dystonia remains largely unknown. Thalamotomy and pallidotomy have been reported to be useful for controlling dystonia. There is, however, a risk of complications with bilateral thalamotomy and pallidotomy. In addition, it is uncertain whether the therapeutic effect of thalamotomy and pallidotomy lasts sufficiently long to justify such irreversible ablative surgery. Recently, deep brain stimulation (DBS) has been reported as an effective therapy for movement disorders [4–6, 8, 13]. It yields therapeutic effects similar to those produced by thalamotomy and pallidotomy, and is associated with a smaller risk of complications. Furthermore, DBS is basically a reversible procedure. On the basis of these advantages, we have been employing DBS as a therapy for primary generalized dystonia. In this report, we summarize our experience of DBS of the globus pallidus internus (GPi) in patients with primary generalized dystonia.

Patients and methods

We have so far undertaken GPi-DBS in 5 patients with primary generalized dystonia. They comprised 3 males and 2 females. They had no history of neuroleptic exposure, detectable brain damage or family history of dystonia. The age at onset of dystonia ranged from 8 to 45 years and the age at surgery for GPi-DBS ranged from 17 to 59 years. Two of the patients had been treated previously by bilateral thalamotomy or unilateral pallidotomy at other clinics. Magnetic resonance imaging (MRI) revealed no lesions other than the lesions caused by such ablative surgery in each of the patients. They underwent GPi-DBS since new symptoms or recurrence of their previous condition had developed after the above ablative surgeries. The inclusion criteria for the DBS trial on dystonia were that the patient was diagnosed as having dystonia which was refractory to any medication. The exclusion criteria included significant cognitive dysfunction, active psychiatric symptoms, and evidence of other central nervous system disease or other systemic medical disorders. All patients underwent bilateral stimulation.

The methods of MRI and microelectrode-guided stereotaxy, and electrode implantation were similar to those reported previously [4]. No sedation was employed during the surgery. The boundaries of the GPi were located by MRI and verified by the recording of spontaneous neuronal activity using tungsten semimicroelectrodes (impedance: 0.5 Mohm). The visual evoked potential was also monitored to avoid injury of the optic tract. The trajectory of the semimicroelectrode was directed from the frontal burr hole at an angle of 50–60 degrees from the horizontal plane. A DBS electrode (Medtronic; Minneapolis, MN) was inserted into the trajectory, which was confirmed to be appropriate. The DBS electrode, which has 4 contact points numbered 0–3 sequentially from the most distal contact (0) to the most proximal contact (3), was placed in such a way that contact 0 was located in the posteroventral portion of the GPi. The contact 0 was located at 3–6 mm below the midpoint between the anterior and posterior commissures, and 19–21 mm lateral to the midline. Each contact of the DBS electrode was 1.5 mm long, and the contacts were 1.5 mm apart from one another. High-resolution MRI was employed for confirmation of the locations of the implanted DBS electrodes postoperatively. When DBS was found to be useful during the test stimulation period for a week, an implantable pulse generator (Itrel 2, Medtronic; Minneapolis, MN) was implanted into the subclavian region and connected to the DBS electrode. The stimulation parameters and contact points used for the GPi-DBS were modified at each follow-up visit to our clinic, based on the results of neurological examinations as well as the patient's report concerning his or her activity of daily living.

The symptoms of dystonia were scored by the Burke-Fahn-Marsden dystonia rating scale (BFMDRS; maximum = 120). In addition, the activity patterns of selected muscle groups were studied by surface electromyography at rest or during execution of simple motor tasks. These clinical evaluations were performed in each patient before surgery, and at 6 months after surgery.

Results

No surgical complications were encountered in the present series. An improvement in dystonia was observed soon after the initiation of GPi-DBS. Relief from dystonia-associated pain was another benefit afforded to the patients. During a period of months or even years after surgery, additional progressive improvement was noted. The effect of GPi-DBS continued to be seen for hours or days after turning the stimulation off. The length of time required for the symptoms of dystonia to return to the original level after temporary termination of the GPi-DBS increased progressively after the initiation of GPi-DBS. The score on the BFMDRS ranged from 18 to 62 before surgery (Table 1). The score at 6 months after surgery reached a level ranging from 4 to 23. The improvement in score ranged from −51% to −92%. These levels of improvement have been maintained so far during the entire follow-up period ranging from 6 months to 2 years. GPi-DBS produced a marked effect even in patients who had previously undergone thalamotomy or pallidotomy.

At 6 months after surgery, all patients were receiving bipolar stimulation with a wide interpolar distance, using contact 0 or 1 as the cathode and contact 2 or 3 as the anode. Stimulation was being performed at an intensity of around 2.0 V with a pulse width of 0.21 ms at a high frequency ranging from 120 to 140 Hz. All

Table 1. *Changes in scores on the Burke-Fahn-Marsden dystonia rating Scale (BFMDRS) by GPi-DBS at 6 months after surgery*

Case	Sex	Age (yr)		BFMDRS		Improvement by GPi-DBS
		at onset	at surgery	before	at 6 mos	
1	M	14	20	56	16	−71%
2	M	15	17	50	4	−92%
3	F	22	59	18	8	−56%
4	F	8	32	62	23	−63%
5	M	45	46	41	20	−51%

patients judged that these stimulation procedures produced the best effect among the various combinations of stimulation parameters and conditions they had experienced. No stimulation-related side effects were induced at the stimulation intensity required to achieve these effects.

Illustrative case

A right-handed 17-year-old male (Case 2) experienced gradual onset and worsening of generalized dystonia, beginning at age 15. He was otherwise in good health. There was no history of neuroleptic exposure and no family history of dystonia. Neuroradiological examinations including MRI yields normal findings. His involuntary movements and abnormal posture were resistant to medical treatment. His level of consciousness and cognitive function were completely normal. There was no motor palsy and no obvious sensory disturbance. Continuous dystonic movements involving mainly the trunk were noted. He bent his trunk forwards completely and showed a dystonic posture and tremor of the left foot. When walking, action-induced bending of the trunk markedly interfered with his gait. Electromyography before surgery demonstrated highly abnormal contraction of various muscles, especially the rectus abdominis muscle. His score on the BFMDRS was 50 before surgery.

DBS electrodes were implanted, placing contact 0 at 5 mm below the midpoint between the anterior and posterior commissures, and 20 mm lateral to the midline. No surgical complications were encountered. A dramatic reduction in muscular tone of the patient's trunk and neck was noted immediately after initiation of the GPi-DBS. The abnormal contraction observed in various muscles disappeared. During a period of several months after surgery, additional progressive improvement was noted. The maximum improvement was reached at 6 months after surgery. His score on the BFMDRS was 4 at 6 months after surgery. No stimulation-related side effects were induced at the stimulation intensity required for this effect. The maximum improvement has so far continued for 2 years.

Discussion

Effects of GPi-DBS on various forms of dystonia, including primary or secondary as well as generalized or segmental varieties, have been reported previously by several authors as case reports or in series with small numbers of patients. Concerning GPi-DBS for primary generalized dystonia, the improvement in score on the BFMDRS reported in the literature ranged from −14% to −90%. Our experience is consistent with these findings. It has been suggested that GPi-DBS tends to provide better results in early onset of primary generalized dystonia than in other cases. An excellent effect was reported especially in DYT1 [2, 11]. The present study, in agreement with previous reports, demonstrated that progressive improvement continued for a period of months or even years, and the length of time required for the symptoms of dystonia to return to the original level after temporary termination of the GPi-DBS increased progressively [5, 12]. We also found that GPi-DBS produces a marked effect even in patients who had previously undergone thalamotomy or pallidotomy and developed new symptoms or recurrence.

Our experience suggests that the best effect in each patient could be achieved by bipolar stimulation with a wide interpolar distance. However, Krauss *et al.* and Roubertie *et al.* employed monopolar stimulation [5, 11]. The pulse width (0.21 ms) and frequency of stimulation (130–160 Hz) used by these investigators were comparable to those employed in the present study (0.21 ms and 120–140 Hz). On the other hand, Krauss *et al.* [5] did use a higher intensity (3.1–5.0 V) than that employed by us in the present study (2.0 V). Kumar *et al.* [6] reported that bipolar stimulation at a high intensity (10 V) with a wide pulse width (1.0 ms) and relatively low frequency (50 Hz) provided better effects on symptoms of dystonia. They demonstrated by positron emission tomography that GPi-DBS could attenuate excessive neuronal activity in the primary and association motor cortex. Although a conclusion to these disparities in stimulation procedures must await further clinical experience, it does appear that modification of neural activities over wide areas within the GPi by stimulation at a relatively high intensity or multiple sites is required to control primary generalized dystonia.

The improvement in the BFMDRS achieved by GPi-DBS in primary generalized dystonia is apparently to be comparable to that caused by pallidotomy, which reportedly ranges from −65% to −80%. Pallidotomy produces a better effect on early onset primary generalized dystonia by early surgery, similarly to GPi-DBS [3, 10, 14]. Progressive improvement for at least 6 months has also been reported for pallidotomy [1, 7, 9]. One advantage of GPi-DBS over pallidotomy may be a smaller risk of complications. In addition, it has been reported that new symptoms or recurrence of the previous condition may frequently develop through disease progression after ablative surgery [1]. Little is yet known regarding the long-term effects of GPi-DBS in primary generalized dystonia. Nevertheless, GPi-DBS appears to have an advantage over pallidotomy, in that DBS does not produce irreversible lesions.

Conclusion

GPi-DBS is an important therapeutic option in many patients with primary generalized dystonia. Further studies are needed to determine the optimal stimulation procedures and the long-term effects. GPi-DBS can achieve a marked effect in patients who have developed new symptoms or recurrence after ablative surgery.

Acknowledgment

This work was supported by grants from the Ministry of Education, Science and Culture of Japan (B09470302 and A12307029).

References

1. Cooper IS (1976) 20-Year follow up study of the neurosurgical treatment of dystonia musculorum deformans. Adv Neurol 14: 423–452
2. Coubes P, Roubertie A, Vayssieres N, Hemm S, Echenne B (2000) Treatment of DYT1-generalised dystonia by stimulation of the internal globus pallidus. Lancet 355: 2220
3. Iacono RP, Kuniyoshi SM, Lonser PR, Maeda G, Inae AM, Ashwal S (1996) Simultaneous bilateral pallidoansotomy for idiopathic dystonia musculorum deformans. Pediatr Neurol 14: 145–148
4. Katayama Y, Oshima H, Fukaya C, Kawamata T, Yamamoto T (2001) Control of post-stroke movement disorders using

chronic motor cortex stimulation. Acta Neurochir (Wien) [Suppl] 79: 89–92
5. Krauss JK, Pohle T, Weber S, Ozdoba C, Burgunder JM (1999) Bilateral stimulation of globus pallidus internus for treatment of cervical dystonia. Lancet 354: 837–838
6. Kumar R, Dagher A, Hutchison WD, Lang AE, Lozano AM (1999) Globus pallidus deep brain stimulation for generalized dystonia: clinical and PET investigation. Neurology 53: 871–874
7. Lin JJ, Lin GY, Shih C, Lin SZ, Chang DC, Lee CC (1999) Benefit of bilateral pallidotomy in the treatment of generalized dystonia. J Neurosurg 90: 974–976
8. Loher TJ, Hasdemir MG, Burgunder JM, Krauss JK (2000) Long-term follow-up study of chronic globus pallidus internus stimulation for posttraumatic hemidystonia: case report. J Neurosurg 92: 457–460
9. Lozano AM, Kumar R, Gross RE, Giladi N, Hutchinson WD, Dostrovsky JO, Lang AE (1997) Globus pallidus internus pallidotomy for generalized dystonia. Mov Disord 12: 865–870
10. Ondo WG, Desaloms JM, Jankovic J, Grossmann RG (1998) Pallidotomy for generalized dystonia. Mov Disord 13: 693–698
11. Roubertie A, Echenne B, Cif L, Vayssiere N, Hemm S, Coubes P (2000) Treatment of early-onset dystonia: update and a new perspective. Childs Nerv Syst 16: 334–340
12. Tronnier VM, Fogel W (2000) Pallidal stimulation for generalized dystonia: report of three cases. J Neurosurg 92: 453–456
13. Vesper J, Klostermann F, Funk TH, Stockhammer F, Brock M (2001) Deep brain stimulation of the globus pallidus internus (GPI) for torsion dystonia: a report of two cases. Acta Neurochir (Wien) [Suppl] 79: 83–88
14. Vitek JL, Zangh J, Evatt M, Mewes K, Delong MR, Hashimoto T, Triche S, Bakay RAE (1998) GPi pallidotomy for dystonia: outcome and neuronal activity. Adv Neurol 78: 211–219

Correspondence: Yoichi Katayama M.D., Ph.D. Department of Neurological Surgery and Division of Applied System Neuroscience, Nihon University School of Medicine, 30-1 Ohyaguchi Kamimachi, Itabashi-ku, Tokyo 173-8610, Japan. E-mail: ykatayam@med.nihon-u.ac.jp

Acta Neurochir (2003) [Suppl] 87: 129–131

Neurosurgical treatment for writer's cramp

T. Taira, S. Harashima, and **T. Hori**

Department of Neurosurgery, Neurological Institute, Tokyo Women's Medical University, Tokyo, Japan

Summary

Background. Writer's cramp is a type of focal dystonia due to dysfunction of the pallido-thalamo-cortical circuit. The symptom is refractory to most conservative treatment, though botulinum toxin injection is generally used for symptomatic relief. As a surgical treatment of dystonia we performed stereotactic nucleus ventrooralis (Vo) thalamotomy for dystonic cramp of the hand.

Method. Eight patients (5 men, 3 women, age 26–40 yrs, mean 32.1 yrs) with medically intractable task-specific focal dystonia of the hand underwent Vo thalamotomy. Stereotactic target was chosen at the junction of the anterior and posterior Vo nuclei.

Findings. The mean duration of the symptom ranged from 3 to 6 years (mean, 4.0 yrs). All patients had complained of difficulty in writing. Six patients were professional workers, such as comic artist, guitarist, and barber, and, because of the dystonic symptoms at their professional work, they had stopped pursuing their profession. All patients showed immediate postoperative disappearance of dystonic symptoms, and the effect was sustained during the follow up period (3–29 months, mean 13.1 mo) except in one case. One patient showed partial recurrence of the symptom and underwent second thalamotomy 5 months after the initial surgery with satisfactory results. The score of the writer's cramp rating scale significantly ($p < 0.001$) decreased after Vo thalamotomy. There was no permanent operative complication. There was no mortality or permanent morbidity.

Interpretation. Although a longer follow-up is needed, stereotactic Vo thalamotomy is a useful and safe therapeutic option for writer's cramp.

Keywords: Focal dystonia; writer's cramp; nucleus ventrooralis; thalamotomy.

Introduction

Writer's cramp is a disabling symptom of slow, untidy script, excessively tight grip of the pen, and/or forearm and hand tension and used to be considered a disturbance of psychogenic origin. However, since Sheehy and Marsden [9] proposed that writer's cramp is a physical dystonic disorder and introduced a concept of focal dystonia of the hand, a sufficient number of reports have been supporting this proposal [4, 5, 11].

Writer's cramp is the most recalcitrant to existing treatment [4, 11]. It has been reported that surgical treatment has no role in management of writer's cramp [9], while there have been several case reports suggesting that thalamotomy effectively alleviated the symptoms [1–3, 7, 8, 10]. Recent studies on the mechanisms of dystonia indicate that changes in neuronal activity that begin in the basal ganglia produce changes in neuronal function throughout the pallido-thalamo-cortical motor loop [11]. Improvement of dystonic symptoms can be expected by interrupting this loop at a certain anatomical point. We report our results of Vo thalamotomy for writer's cramp.

Patients and methods

Between September 1999 and January 2002, eight patients (5 men, 3 women, age 26–40 yrs, mean 32.1 yrs) with medically refractory dystonic cramp of the hand underwent Vo thalamotomy. The diagnosis of focal hand dystonia was made with careful history taking and physical examination. We used the methods described by Wissel *et al.* [12] for objective assessment.

Surgical procedure

Under local anesthesia, the Leksell stereotactic frame (Eleka model G, Stockholm, Sweden) was fixed to the patient's skull. Contiguous 1 mm slice T1 weighted MR images and 1 mm slice computer tomographic (CT) scan images were acquired. The tentative stereotactic target was chosen at 2 mm posterior to the mid-intercommissural point, 1 mm dorsal to the intercommissural line, and 13.5 mm lateral from the midline. This point corresponds to the junction between the nuclei of anterior and posterior Vo of the thalamus. All the operative procedures were performed purely under local anesthesia. We placed the patient in a semi-sitting position so that the patient was able to perform writing or other dystonia-evoking activities. From a frontal burr hole, a monopolar radiofrequency probe (1.0 mm diameter tip with 2.0 mm uninsulated length) was inserted. Electrical stimulation (145 Hz, 100 μsec pulse width, up to 10 mA) was performed before lesioning to avoid surrounding structures. We carefully checked the symptoms and side effects with a trial lesion at 45 °C. If the dystonic symptoms im-

Table 1. *Number of patients, penetration, and recorded cells in the present study*

	No. of patients	No. of penetration	No. of cells
ET	14	23	878
PD	17	31	780

entially to passive movements of a joint were classified as kinesthetic (Ki) neurons. The mean firing rate, the interspike interval (ISI), the autocorrelogram and the discharge pattern for each neuron were investigated. The ISIs allowed an evaluation of the degree of bursting of the units, following an algorithm described by Hutchison *et al.* [11]. If a unit had an index of more than 10 from the reciprocal of the modal interval divided by the mean firing rate, it was regarded as bursting. The autocorrelogram evaluated the rhythmicity and the cut-off point for the rhythmicity was 0.95. A bursting unit was classified as tremor cell if the frequency of clearly identifiable peaks on the autocorrelogram corresponded to that of tremor measured by the EMG from the arm or leg. The differences between each group were statistically evaluated using the unpaired t-test with significance level of 0.05.

Results

Comparison of mean firing rates of ET patients and PD patients

A total of 1658 units were isolated along the 54 penetrations performed in 14 ET patients and 17 PD patients (Table 1). Somatosensory cells were often found in the Vim and Vop of ET patients as in the PD patients. The Ki neurons were mainly found in the Vim (47.41%), whereas the Vo neurons were predominant in the Vop (71.43%) in ET patients. As reported previously, the mean firing rates of the ET patients were higher than those of the PD patients in both the Vop and Vim nuclei ($p < 0.05$) (Fig. 1). In addition, the mean firing rates of the Ki neurons of the ET patients were higher than those of the PD patients in the Vim nuclei ($p < 0.05$) (Fig. 2). However, the mean firing rates of the ventralis caudalis (Vc) neurons, which respond to sensory stimulation, were similar in each group (Fig. 3).

The firing pattern of Vim and Vop nuclei

An analysis of the firing patterns revealed some significant differences between the ET and PD patients for the Vo or Ki neurons. The analysis of the incidence of the bursting neurons revealed that the Vop nucleus of the ET patients had much lower number of bursting

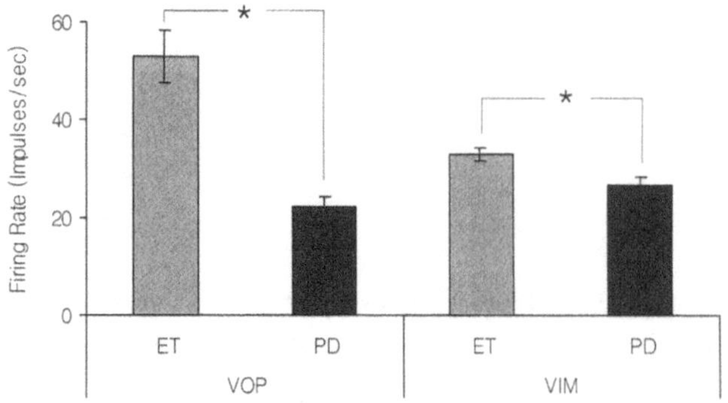

Fig. 1. Mean firing rates of the nucleus ventralis oralis posterior (*VOP*) and the nucleus ventralis intermediates (*VIM*) neurons in an essential tremor (*ET*) and Parkinson's disease (*PD*) patients. There was a significant difference between mean firing rates of the VOP and VIM neurons in ET and PD patients (*$p < 0.05$)

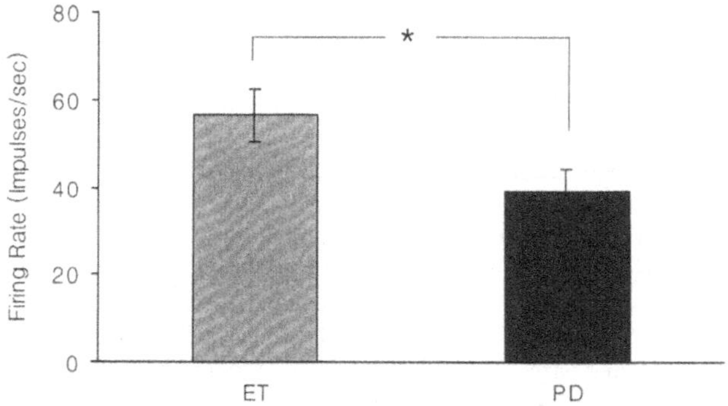

Fig. 2. Mean firing rates of the kinetic (*Ki*) neurons in the nucleus ventralis intermediates (*VIM*) of essential tremor (*ET*) and Parkinson's disease (*PD*) patients. There was significant difference between the mean firing rates of the Ki neurons in the VIM of ET and PD patients (*$p < 0.05$)

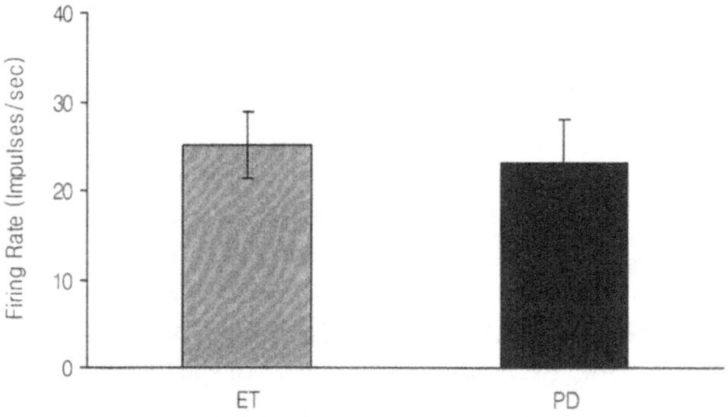

Fig. 3. Mean firing rates of the somatosensory neurons in the ventralis caudalis (*VC*) of essential tremor (*ET*) and Parkinson's disease (*PD*) patients. There was no significant difference between the mean firing rates of the somatosensory neurons in the VC of ET and PD patients ($p > 0.05$)

neurons (9.58%) than the PD patients (33.82%). However, in the Vim nucleus, both groups possessed bursting neurons even though the incidence was slightly different (ET patients, 26.25%; PD patient, 38.91%). Tremor cells were less frequently observed in the ET patients at the Vim nuclei compared to those of the PD patients. Interestingly, tremor cells were not found in the Vop nucleus of the ET patients.

Discussion

An essential tremor is supposed to be generated within the olivocerebellar circuits. This contrasts with a resting tremor of the PD, which is probably generated in the basal ganglia loop [3]. Among the most convincing clinical evidence, is the observation that an ET disappears after the lesions of the cerebellum, the pons, or the thalamus, each of which forms a part of a cerebellocortical loop [5, 6, 14]. It has been shown that in different species, cells of the inferior olive are synchronized and that rhythmic activity is transferred through the cerebellum and the reticulospinal projection to the motoneurons [16].

However, the basic physiological mechanism of tremor genesis in PD is more complicated than that in ET. It is generally agreed that a resting tremor is produced within the basal ganglia loop, but the location and physiology of the related tremor genesis are still a matter of debate. It has recently been suggested that a resting tremor is produced in the cerebellothalamic loop. This is because the cerebellum shows prominent hyperactivity in a tremor, and a PD tremor can be successfully treated by the placement of a lesion in the Vim nucleus of the thalamus, which is the target of the cerebellothalamic fibers [5, 15]. This hypothesis implies that a resting tremor should cease when the cerebellum is removed in a patient with a PD tremor. However there was a report of a continuous tremor after surgery, which contrasts with the cerebellar hypothesis for tremor genesis [2]. Certainly, a single unit recording in patients during stereotactic surgery has revealed that the number of neurons with a clear relationship with a peripheral tremor varies according to the different nuclei, with the majority being found in the motor thalamus, followed by the globus pallidus internus (GPi) and the subthalamic nucleus (STN). In addition, other possibilities, such as a corticosubthalamic-internal pallidum-thalamocortical loop, may be equally likely to form an oscillating loop [3].

It is generally believed that the Vo neurons are located mainly in the pallidal-receiving area of the motor thalamus and the Ki neurons are found in the cerebellar receiving area of the thalamus [4]. In this category, our results are consistent with these findings in that an increasing inhibitory input from the GPi should decrease the firing of the Vo and Ki neurons in PD patients. Moreover, it is possible that the cerebellar hyperactivity can account for the increase in the spontaneous firing rates of the Vo and Ki neurons in ET patients, as revealed using the microelectrode recording techniques of this study. Since the cerebellar projections to the thalamus are excitatory, the increased excitability of the Vim neurons might lead to increased kinesthetic feedback and contribute to the generation of a tremor [3].

The lack of a significant increase in bursting activity in the pallidal receiving areas in PD patients has been reported previously by Dostrovsky *et al.* [4]. In addition, they suggested that bursting is unlikely to be an important factor in the etiology of a parkinsonian tremor. Interestingly, the number of tremor cells in the motor thalamus in ET patients (21/339 units, 6.19% for Vim) in this study was much lower than those in PD patients (34/257 units, 13.23% for Vim). These findings may be consistent with the observation in PD patients reported by Dostrovsky *et al.* [4]. Recently, Halliday *et al.* [8] demonstrated no correlation between the tremor signal and the low-frequency activity recorded from the primary motor cortex in individuals with an essential tremor. They suggested that the tremor originated through the descending pathways rather than in the primary motor cortex.

In contrast, stereotactic operations in humans and the increased activity in the motor thalamus, as revealed in this study, indicate that the thalamus plays an important role in the generation of an essential tremor. This suggests that considering the strong thalamocortical projections present, there might also be some cortical involvement [3, 5, 9]. However, there has been no direct proof of a cerebral cortex contribution to the central oscillator that produces the essential tremor.

Conclusions

This study demonstrated the different pathophysiological mechanisms of ET and compared it with PD. This might provide a rationale for stereotactic surgery on Vim neurons in ET patients. However, the complex pathophysiological mechanisms leading to the clinical expression of an essential tremor are still unclear.

Acknowledgment

This work was supported by grant No. (2000-2-20600-001-3) from the Basic Research Program of the Korea Science & Engineering Foundation.

References

1. Chang JW, Lee BH, Lee MS, Chang JH, Park YG, Chung SS (2001) Microelectrode recording-guided deep brain stimulation in patients with movement disorders (The first trial in Korea). In: Lultas-Ilinsky K, Ilinsky IA (eds) Basal ganglia and thalamus in health and movement disorders. Kluwer Academic/Plenum Publishers, New York, pp 341–347
2. Deuschl G, Wilms H, Krack P, Wurker M, Heiss WD (1999) Function of the cerebellum in parkinsonian rest tremor and Holmes' tremor. Ann Neurol 46: 126–128
3. Deuschl G, Raethjen J, Lindehann M, Krack P (2001) The pathophysiology of tremor. Muscle Nerve 24: 716–735
4. Dostrovsky JO, Molnar GF, Pilliar A, Hutchison WD, Davis KD, Lozano AM (2001) Neuronal activity in motor thalamus of Parkinson's disease patients. In: Lultas-Ilinsky K, Ilinsky IA (eds) Basal ganglia and thalamus in health and movement disorders. Kluwer Academic/Plenum Publishers, New York, pp 267–274
5. Duncan R, Bone I, Melville ID (1988) Essential tremor cured by infarction adjacent to the thalamus. J Neurol Neurosurg Psychiatry 51: 591–592
6. Dupuis MJ, Delwaide PJ, Boucquery D, Gonsette RE (1989) Homolateral disappearance of essential tremor after cerebellar stroke. Mov Disord 4: 183–187
7. Elble RJ (1996) Central mechanisms of tremor. J Clin Neurophysiol 13: 133–144
8. Halliday DM, Conway BA, Farmer SF, Russell AJC, Rosenberg JR (2000) Coherence between low-frequency activation of the motor cortex and tremor in patients with essential tremor. Lancet 355: 1149–1153
9. Hellwig B, Haubler S, Schelter B, Lauk M, Guschlbauer B, Timmer J, Lucking CH (2001) Tremor-correlated cortical activity in essential tremor. Lancet 357: 519–523
10. Hua SE, Lenz FA, Zirch TA, Reich SG, Dougherty PM (1998) Thalamic neuronal activity correlated with essential tremor. J Neurol Neurosurg Psychiatry 64: 273–276
11. Hutchinson WD, Levy R, Dostrovsky JO, Lozano AM, Lang AE (1997) Effects of apomorphine on globus pallidus neurons in parkinsonian patients. Annals Neurol 42: 767–775
12. Llnias R, Volkind RA (1973) The olivo-cerebellar system: functional properties as revealed by harmaline-induced tremor. Exp Brain Res 18: 69–87
13. Louis ED (2001) Essential tremor. N Eng J Med 345: 887–891
14. Nagaratnam N, Kalaabail G (1997) Contralateral abolition of essential tremor following a pontine stroke. J Neurol Sci 149: 195–196
15. Pahwa R, Lyons K, Koller W (2000) Surgical treatment of essential tremor. Neurology 54 [Suppl] 4: s39–s44
16. Wilm H, Sievers J, Deuschl G (1999) Animal models of tremor. Mov Disord 14: 557–571

Correspondence: Jin Woo Chang, M.D., Ph.D., Department of Neurosurgery, Yonsei University College of Medicine, CPO Box 8044, Seoul, 120-752, Korea. E-mail: jchang@yumc.yonsei.ac.kr

Acta Neurochir (2003) [Suppl] 87: 137–139

Localization of thalamic cells with tremor-frequency activity in Parkinson's disease and essential tremor

K. Kobayashi[1], **Y. Katayama**[1,2], **M. Kasai**[1], **H. Oshima**[1], **C. Fukaya**[1,2], and **T. Yamamoto**[1,2]

[1] Department Neurological Surgery, Nihon University School of Medicine, Tokyo, Japan
[2] Division of Applied System Neuroscience, Nihon University School of Medicine, Tokyo, Japan

Summary

It has been reported that parkinsonian and essential tremor can be controlled by deep brain stimulation or radiofrequency lesion within the cluster of cells with a tremor-frequency activity in the ventral thalamic nuclei. However, there have been very few reports about the exact localization of cells with tremor-frequency activity in the ventral thalamic nuclei. In the present study, we investigated the localization of cells with tremor-frequency activity in the ventral thalamic nuclei employing autopower spectrum and coherence analysis. Activity of a total of 130 cells, 63 in patients with parkinsonian tremor and 67 in patients with essential tremor, were recorded from the area anterior to the nucleus ventralis caudalis. Among these cells, 31 cells showed a coherence of greater than 0.4 to the electromyographic activity of both agonist and antagonist muscles. The proportion of cells exhibiting tremor-frequency activity were 26.8% in the nucleus ventralis intermedius (Vim) and 25.0% in the nuclei ventralis oralis posterior et anterior (Vop + Voa). There were no significant differences in proportion by nuclear location or disease. The present study demonstrated that cells with tremor-frequency activity are widely distributed over the area extending from the Vim to the Vop + Voa. This indicates that the best location for placeing electrodes for deep brain stimulation or a radiofrequency lesion cannot be defined by identification of cells with tremor-frequency activity alone.

Keywords: Brain stimulation; essential tremor; involuntary movement; Parkinson's disease; thalamus; tremor.

Introduction

Cells in the ventral thalamic nuclei demonstrate activity at frequencies close to that of tremor [1, 2, 5, 7, 12, 13]. Tremor can be abolished when a radiofrequency lesion is centered within the cluster of cells exhibiting such tremor-frequency activity [4, 11, 14, 16]. This observation has led to the speculation that cells with tremor-frequency activity may represent a tremorogenic central pacemaker [5, 6, 13] as well as the idea that identification of the cluster of cells with tremor-frequency activity is an important surgical procedure for appropriate placement of radiofrequency lesions or deep brain stimulation (DBS) electrodes for tremor control. However, there have been very few reports on the localization of cells with tremor-frequency activity in the ventral thalamic nuclei. In the present study, we investigated the localization of cells with tremor-frequency activity in the ventral thalamic nuclei employing autopower spectrum and coherence analysis.

Materials and methods

We analyzed data obtained from the implantation of a total of 8 DBS electrodes in 7 cases of parkinsonian or essential tremor. Extracellular signals were recorded with a tungsten microelectrode (diameter of exposed tip, approximately 10 μm; impedance, 0.2–0.5 Mohm at 1000 Hz), and the filter conditions were −6 dB at 300 and 10 kHz. The spike train of isolated cells within the thalamus was discriminated with a level detector. The first trajectory was directed towards the anterior aspect of the posterior commissure (PC) on the level of the anterior commissure – PC line at 15 mm lateral to the midline. Cells responding to a passive joint movement of the contralateral limbs without a response to skin deformation caused by these stimuli (deep sensory cells) and those responding to light touch on the skin of the face and contralateral limbs (cutaneous sensory cells) appeared on this trajectory. We defined the nucleus ventralis intermedius (Vim)-nucleus ventralis caudalis (Vc) border as a vertical line, at which the most anterior deep sensory cells responding briskly to a passive joint movement were encountered in the initial trajectory. Cutaneous sensory cells were usually observed just posterior to this area. Microstimulation produced paresthesia in the area in which deep and cutaneous sensory cells were found. The Vim-Vc border as determined by these observations was located 1.9–5.5 mm anterior to the anterior aspect of the PC. In each cell, the distance from the neurophysiologically defined Vim-Vc border and the distance from the midline were determined based on the intraoperative roentgenograms.

Electromyographic (EMG) activity was monitored from up to 8 muscles including the thenar muscles, flexor digitorum superficialis, extensor digitorum communis, flexor and extensor carpi ulnaris, biceps, triceps and deltoid. The EMG activity and spike trains were digitized. A fast Fourier transform (FFT) was taken of both the

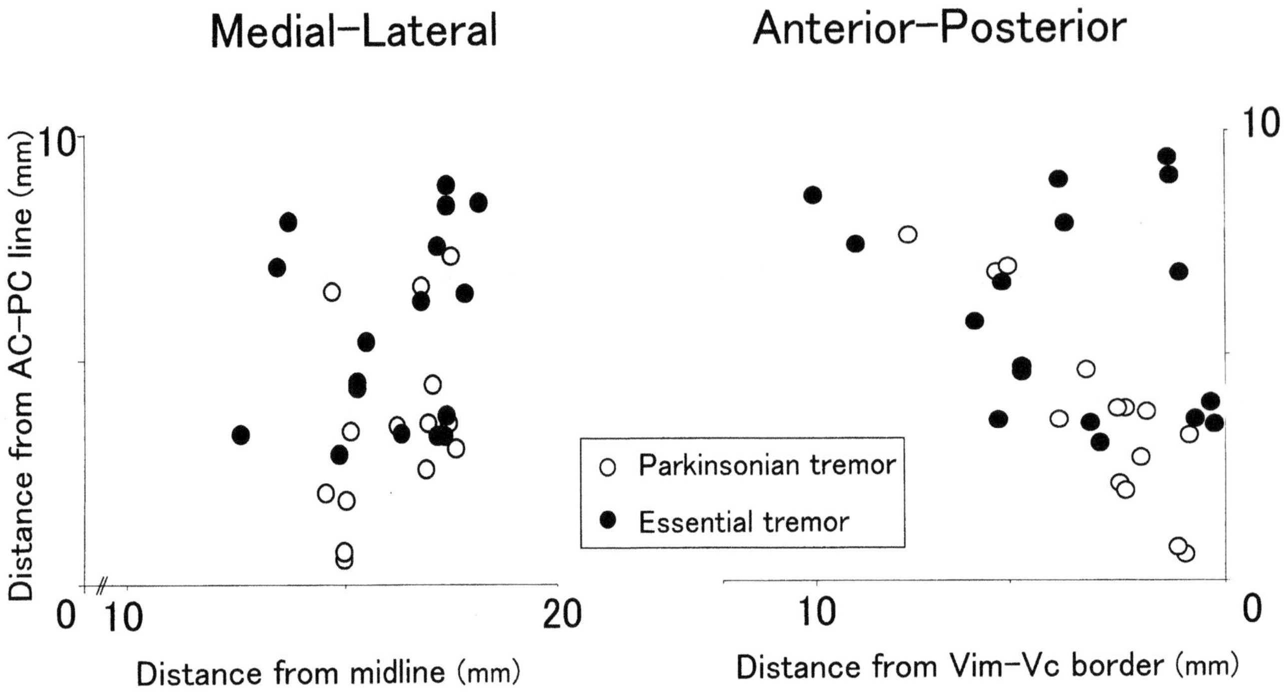

Fig. 1. Localization of cells with tremor-frequency activity (coherence of greater than 0.4). Left, Coronal plane (abscissa and ordinate, distance from the midline and the AC-PC line, respectively); right, sagittal plane (abscissa and ordinate, distance from the Vim-Vc border and the AC-PC line, respectively). A total of 31 cells were recorded at the Vim (n = 15), Vop + Voa (n = 12), and anterior to the Voa (n = 4). *AC* Anterior commissure; *PC* posterior commissure

spike trains discriminated with a level detector and the EMG activity to determine the spike and raw EMG spectra. The autopower spectrum measures the intensity of the signal as a function of frequency. The coherence is used to estimate the probability that two signals are correlated at a given frequency [3, 7, 10].

Results

A total of 130 cells, 63 in patients with parkinsonian tremor and 67 in patients with essential tremor, located anterior to the Vim-Vc border were investigated. The activity of cells located posterior to the Vim-Vc border was not analyzed. The frequency of the EMG activity of tremors usually varies among different body parts. We therefore first determined the EMG activity having the highest level of coherence with the spike spectrum at tremor frequency. Since tremor represents the alternating contraction of agonist and antagonist muscles, it is not known whether a given cell activity is physiologically related to the EMG activity with the highest level of coherence or the EMG activity of its antagonist muscle. Among the cells examined, 31 cells showed a coherence of greater than 0.4 to the EMG activities of both agonist and antagonist muscles, which has commonly been employed as a criterion of significance [3].

The proportion of cells exhibiting tremor-frequency activity was analyzed in relationship to nuclear location and disease. The location of these cells was expressed with reference to the Vim-Vc border (Fig. 1). They were distributed over an area 0.2–9.8 mm anterior to the Vim-Vc border. The proportion of cells exhibiting tremor-frequency activity were 26.8% in the Vim and 25.0% in the nuclei ventralis oralis posterior et anterior (Vop + Voa). The results demonstrated no significant difference (chi-square test) in proportion by nuclear location or disease (Table 1).

Table 1. *Location and numbers of cells showing tremor-frequency activity*

Disease	Cells with tremor-frequency activity/Cells examined			
	Nuclear location			
	Anterior to Voa	Vop + Voa	Vim	Total
Parkinsonian tremor (n = 63)	1/10 (10.0%)	6/25 (24.0%)	7/28 (25.0%)	14/63 (22.2%)
Essential tremor (n = 67)	3/16 (18.8%)	6/23 (26.1%)	8/28 (28.6%)	17/67 (25.4%)
Total (n = 130)	4/26 (15.4%)	12/48 (25.0%)	15/56 (26.8%)	31/130 (23.8%)

Discussion

Ohye *et al.* [15] reported that cells with tremor-frequency activity responding to a passive joint movement are located in the lateral portion of the Vim, and a small radiofrequency lesion restricted to this area affords the best effect on tremor. They emphasized that tremor is arrested in an almost point-to-point relation between coagulation of these cells and control of tremor within their receptive field. On the other hand, Lenz *et al.* [8] found that a small lesion within an area where the majority of cells are not responsive to a passive joint movement provided an excellent effect on tremor. The present study demonstrated that cells with tremor-frequency activity are widely distributed over the area extending from the Vim to the Vop + Voa. Thus, the area which provides the best effect on tremor cannot be determined by identification of cells with tremor-frequency activity alone.

In our experience, the effect of thalamic DBS on tremor can be maximized by appropriate orientation of DBS electrode array within the ventral thalamic nuclei. Better effect on tremor is obtained by DBS electrode array passing through both the Vop and Vim, as compared to DBS electrode array passing through the Vim only. This experience is consistent with the present finding that cells with tremor-frequency activity are distributed widely over the area extending from the Vim and the Vop + Voa.

Acknowledgment

This work was supported by grants from the Ministry of Education, Science and Culture of Japan (B09470302 and A12307029).

References

1. Albe-Fessard D (1973) Electrophysiological methods for the identification of thalamic nuclei. Z Neurol 205: 15–28
2. Bates JAV (1969) The significance of tremor phasic units in the human thalamus. In: Gillingham FJ, Donaldson IML (eds) Third Symposium on Parkinson's disease. E and S Livingstone, Edinburgh, pp 118–124
3. Benignus VA (1969) Estimation of the coherence spectrum and its confidence interval using the fast fourier transform. IEEE Trans Audio Electroacoustic 17: 145–150
4. Hirai T, Miyazaki M, Nakajima H, Shibazaki T, Ohye C (1983) The correlation between tremor characteristics and the predicted volume of effective lesions in stereotactic nucleus ventralis intermedius thalamotomy. Brain 106: 1001–1018
5. Jasper HH, Bertrand G (1966) Thalamic units involved in somatic sensation and voluntary and involuntary movements in man. In: Purpura DP, Yahr MD (eds) The thalamus. Columbia University Press, New York, pp 365–390
6. Lamarre Y, Joffroy AJ (1979) Experimental tremor in monkey: activity of thalamic and precentral cortical neurons in the absence of peripheral feedback. Adv Neurol 24: 109–122
7. Lenz FA, Tasker RR, Kwan HC, Schnider S, Kwong R, Murayama Y, Dostrovsky JO, Murphy JT (1988) Single unit analysis of the human ventral thalamic nuclear group: Correlation of thalamic "tremor cells" with the 3–6 Hz component of parkinsonian tremor. J Neurosci 8: 754–764
8. Lenz FA, Kwan HC, Dostrovsky JO, Tasker RR, Murphy JT, Lenz YE (1990) Single unit analysis of the human ventral thalamic nuclear group: activity correlated with movement. Brain 113: 1795–1821
9. Lenz FA, Kwan HC, Martin RL, Tasker RR, Dostrovsky JO, Lenz YE (1994) Single unit analysis of the human ventral thalamic nuclear group: tremor-related activity in functionally identified cells. Brain 117: 531–543
10. Mima T, Hallett M (1999) Corticomuscular coherence: Review. J Clin Neurophysiol 16: 501–511
11. Nagaseki Y, Shibasaki T, Hirai T, Kawashima Y, Hirato M, Wada H, Miyazaki M, Ohye C (1986) Long-term follow-up results of selective VIM-thalamotomy. J Neurosurg 65: 296–302
12. Ohye C, Saito Y, Fukamachi A, Narabayashi H (1974) An analysis of the spontaneous rhythmic and non-rhythmic burst discharge in the human thalamus. J Neurol Sci 22: 245–259
13. Ohye C, Albe-Fessard D (1978) Rhythmic discharges related to tremor in humans and monkeys. In: Chalazonitis N, Boisson M (eds) Abnormal neuronal discharges. Raven Press, New York, pp 37–48
14. Ohye C, Narabayashi H (1979) Physiological study of presumed ventralis intermedius neurons in the human thalamus. J Neurosurg 50: 290–297
15. Ohye C, Shibazaki T, Hirai T, Wada H, Hirato M, Kawashima Y (1989) Further physiological observations on the ventralis intermedius neurons in the human thalamus. J Neurophysiol 61: 488–500
16. Tasker RR, Organ LW, Hawrylyshyn PA (1982) The thalamus and midbrain of man: a physiological atlas using electrical stimulation. Charles C Thomas, Springfield

Correspondence: Kazutaka Kobayashi, M.D., Department of Neurological Surgery, Nihon University School of Medicine, 30-1 Ohyaguchi Kamimachi, Itabashi-ku, Tokyo 173-8610, Japan. E-mail: ittasaitama@hotmail.con

Acta Neurochir (2003) [Suppl] 87: 141–145

MR safety in patients with implanted deep brain stimulation systems (DBS)

A. Sharan[1,2], A. R. Rezai[2], J. A. Nyenhuis[5], G. Hrdlicka[6], J. Tkach[3], K. Baker[4], M. Turbay[2], P. Rugieri[3], M. Phillips[3], and F. G. Shellock[7]

[1] Department of Neurosurgery Thomas Jefferson University, Philadelphia, PA
[2] Section of Stereotactic and Functional Neurosurgery Department of Neurosurgery The Cleveland Clinic Foundation, Cleveland, OH
[3] Department of Radiology The Cleveland Clinic Foundation, Cleveland, OH
[4] Department of Neurology The Cleveland Clinic Foundation, Cleveland, OH
[5] Department of Computer and Electrical Engineering Purdue University, West Lafayette, IN
[6] Medtronic, Minneapolis, MN
[7] Keck School of Medicine, University of Southern, Los Angeles, CA

Summary

Introduction. While it is desirable to perform MRI examinations in patients with deep brain stimulators (DBS), a major safety concern exists regarding the potential for excessive heating secondary to magnetically induced electrical currents. This study was designed to determine the safety of MRI and DBS.

Methods. Standard configurations of DBS systems were tested. In vitro testing was performed using a 1.5-Tesla MR system, a gel-filled phantom, and the body and head RF coils with varying levels of RF energy (SAR). A fluoroptic thermometry system was used to record temperatures.

Results. Using the 1.5-T MRI and body RF transmit coil, the temperature changes ranged from 2.5 to 25.3 °C. Using the 1.5-T MRI and head RF transmit coil, the temperature changes ranged from 2.3 to 7.1 °C.

Conclusions. Excessive heating does occur with certain MR imaging conditions. Under certain conditions determined in this study, patients with DBS may safely undergo anatomical MR imaging. In the future, standardized testing and more comprehensive studies will be needed to ensure the MR safety of neurostimulation systems.

Keywords: Deep brain stimulation; neurostimulation; magnetic resonance imaging; safety; magnetic resonance imaging; implants.

Introduction

Deep brain stimulation (DBS) has rapidly become a commonly utilized treatment for Parkinson's disease, essential tremor, and dystonia. Emerging applications of DBS include the treatment of epilepsy, obsessive compulsive disorder, chronic pain, headaches, and other indications. The number of centers worldwide performing DBS surgery has increased from 50 to over 300 in the past few years. It is clear that there will be increasing medical issues relevant to patients with implanted neurostimulation systems.

Magnetic resonance imaging (MRI) is generally regarded as a safe, noninvasive diagnostic technique and is of critical importance in the management of patients with implanted DBS systems [16, 18]. MRI is necessary in the surgical planning of patients considered for DBS, is necessary for verification of lead position postoperatively, and is needed in cases of poor clinical outcome and/or significant side effects [4, 14, 15, 24]. Further, one must remember that patients who are selected for DBS typically have diseases which are progressive in nature and it is intuitive that these patients may require repeated MRI to evaluate ongoing or changing symptoms or to evaluate other concurrent pathologies such as stroke, neoplasm, hemorrhage, or infectious diseases. Finally, there is increasing enthusiasm in the use of functional MR imaging applied to patients with neurostimulation systems to improve the understanding of the pathophysiology of these neurological diseases as well as unraveling the mechanisms of action.

It is widely believed that patients with implanted neurostimulation systems should generally not be allowed to undergo MR examinations. The potential for serious risks include MR-related heating of electrodes, movement of the leads and neural pulse generators, induced electrical currents, functional disruption of the device, and production of MR related artifacts [2, 4–6,

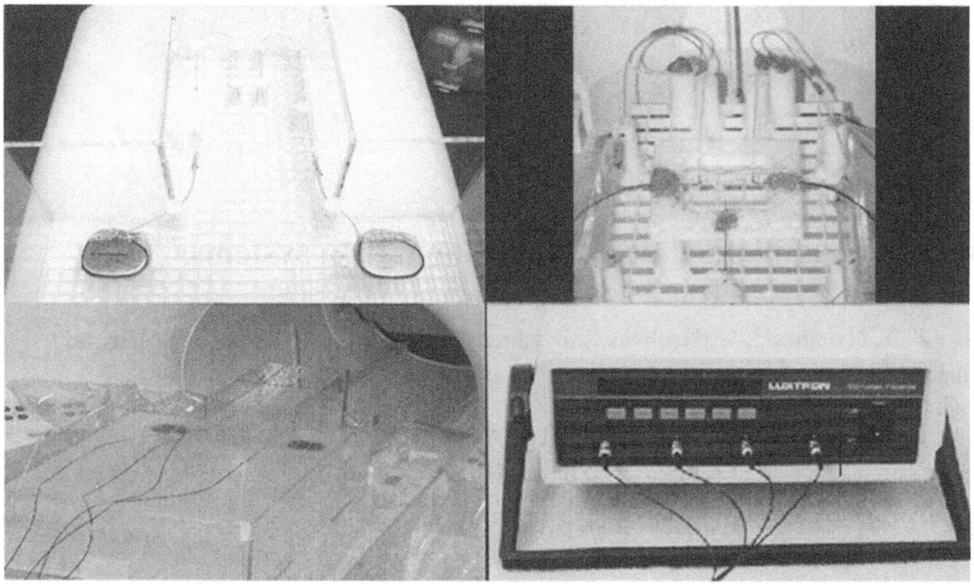

Fig. 1. (a) Plexiglass phantom with plastic frame placed in bottom to hold movable posts for variable positioning of stimulators, leads, extensions and IPG. (b,c) Note bilateral stimulator placement with extensions to bilateral head leads and long cables for thermometry probes in gel filled phantom. (d) Luxtron fiberoptic thermometry system

9, 10, 12–18, 21]. Prior MR safety studies conducted on other electronically-activated devices, such as cardiac pacemakers, have generally recommended strict prohibition of patients with electronically active devices from undergoing MR scanning [20]. Many centers around the world have routinely imaged patients with DBS systems without adverse effects. However, only a limited number of studies have addressed MR safety issues for implantable pulse generators (IPGs) and DBS electrodes [5, 6, 9, 10, 12, 13, 21].

The present report is intended to alert the medical community of the emerging safety issues concerning patients with implanted DBS systems during MR imaging.

Materials and methods

The specific methodology used in the current investigation has been previously described [5, 13]. Briefly, in vitro testing was conducted using a plexiglas phantom designed to approximate the size and shape of the human head and torso was filled with a semi-solid aqueous gel prepared to simulate the electrical conductivity and thermal convection properties of tissue (Figure 1).

The only FDA approved device for chronic deep brain stimulation is the Activa Tremor Control System (Medtronic, Inc. Minneapolis MN). In this study, bilateral DBS electrodes (model 3387, 3389), IPGs (Soletra) were positioned in a manner to emulate the clinical methodology used during implantation of a DBS. The practice of looping the excess electrode around the burr hole was tested as well as testing a scenario where the entire excess length of the electrode and extension was coiled in the pocket of the IPG was conducted.

Temperature monitoring was performed at multiple locations using an MR-compatible fluoroptic thermometry system (Model 790, Luxtron, Santa Clara, CA). MR imaging was performed using a 1.5 Tesla/64 MHz MR system (Vision, software version, Numaris 3.0; Siemens Medical Systems, Iselin, NJ) and either the transmit/receive body RF coil or the transmit/receive head RF coil. Parameters were specifically selected to vary the whole-body and local SARs in these experiments to assess MRI-related heating for the neurostimulators under relatively high and relatively low RF power levels [23].

Results

The results of this study are summarized in Table 1. For the use of the transmit/receive body RF coil, the temperature changes ranged from 2.5 to 25.3 °C. The highest temperature change, 25.3 °C, was recorded with a whole body averaged SAR of 3.90 W/kg.

Table 1. *Results*

Configuration	RF coil	Whole body SAR (W/kg)	MR landmark	Highest temperature change (degree celsius – distal electrode)
No loops	body	3.9	IPG	25.3
Loops	body	3.9	IPG	6.1
Loops	body	0.98	IPG	2.5
Loops	head	0.24	electrode	7.1
Loops	head	0.13	electrode	3.7
Loops	head	0.07	electrode	2.3

For the experiments involving the transmit/receive head RF coil, the temperature changes ranged from 2.3 to 7.1 °C. The highest temperature change, 7.1 °C, was recorded with a whole body averaged SAR of 0.24 W/kg. Few interesting observations should be noted. The temperature increase is extremely rapid during the onset of MR imaging occurring predominantly in the first minute of imaging. Also, a change in electrode configuration resulted in a substantial change in the resultant temperature. Finally, the highest temperature changes were recorded at the distal electrodes in these experiments.

Discussion

Considering the importance of using MR imaging for the management of patients, research studies have been performed to determine risks associated with using MR imaging in the patients with certain electronically-activated implants [16–18]. The MR safety concerns for electrical stimulation devices include heating, magnetic field interactions, induced electrical currents, functional disruption of the operational aspects of these devices, and the generation of MR related artifact [2, 4–6, 9, 10, 12–18, 21].

Heating

The most significant concern with neurostimulation systems is the possibility of generating excessive heating of the components of the neurostimulation systems [16]. Further, it is certain that there are threshold temperatures known to produce reversible (i.e., range 42 to 44 °C), and irreversible thermal lesions (i.e., >45 °C) in the central nervous system [1, 3, 7, 8]. There have been reports of serious brain injuries in two patients with DBS systems, one who received RF diathermy treatments for an unrelated condition, and one who was cardioverted for atrial fibrillation [11, 22]. Thus, there is a realistic potential for injury due to excessive heating of DBS electrodes.

The internalized or externalized lead wire component of the neurostimulation system functions as an antenna in the RF field generated during MR imaging. As a result, an electric current is induced in the wire and as this current is dissipated in the nervous tissue at the electrical contact, there is resultant heating which is proportional to the square of the emanating current. As the specific absorption rate (SAR) is proportional to the square of the induced electric field, the temperature rise at the electrode contacts is proportional to the SAR, as was observed in this study.

Prior investigations

To date, only a small number of investigations have addressed MR safety issues for DBS electrodes, IPG's and other neurostimulation devices with varying results [2, 4–6, 9, 10, 12–18, 21]. As the number of investigations addressing these MR Safety concerns increase, it will be paramount to detail the experimental design to allow for scientific assessment and standardization of the methodology, to allow reproducibility and more importantly, to ensure patient safety. This is largely the reason for variable findings across investigations: (1) the use of bilateral vs. unilateral neurostimulation systems, (2) the different RF energy levels selected for the experiments, (3) the positioning techniques used for the neurostimulation systems (i.e. such assumptions that the lead and/or extension should be arranged in the form a "meander" can not be drawn without conducting specific hypothesis to test these variables), (4) the method used to record temperatures (i.e., fluoroptic thermometry is considered the industry standard), and (5) the use of a phantom filled with gel prepared to simulate the thermal convection properties of human tissue vs. saline. Preliminary experimental work by our group has confirmed that all variable detailed above must be independently investigated.

Present investigations

In early 2001, we initiated a comprehensive set of experiments designed to define MR imaging conditions and parameters that would permit the safe use of this imaging modality in patients with DBS electrodes. In vitro experiments were performed at 1.5-Tesla using a fluoroptic thermometry system to record temperatures at multiple positions for various configurations of IPGs and DBS electrodes placed in a specially-constructed phantom (i.e., used to simulate the human torso) filled with a tissue-equivalent conducting gel. Studies were conducted using both transmit/receive head coils and body coils [5, 13]. Based on the findings reported herein, the following guidelines can be generated for neurostimulators positioned as mentioned in the methodology with a static magnetic field of 1.5 Tesla.

(1) The neurostimulation systems should be interrogated prior to MR imaging to ensure that there

are no malfunction of the electrodes, leads, and extensions and that the amplitude for each neurostimulation system has been set to 0 volts and the system's output should be turned to "off". After MR imaging, the neurostimulation systems should be interrogated to verify functionality and be reprogrammed to stimulation parameters used prior to MR imaging.

(2) If the transmit/receive body RF coil is used for MR imaging, the whole body averaged SAR should not exceed 0.9 W/kg.

(3) If the transmit/receive head RF coil is used for MR imaging, the whole body averaged SAR should not exceed 0.1 W/kg.

(4) MR imaging should be performed using standard techniques that utilize the lowest possible SAR levels, as indicated above.

(5) Patients should be instructed prior to MR imaging to report any unusual sensations that may occur during the examinations and continuously monitored using visual and/or verbal means.

The manufacturer has similarly published guidelines on how patients may safely obtain MR imaging under certain specific scenarios [19].

Conclusions

At this time, more work is needed for a more comprehensive evaluation of heating. These current guidelines are strictly restricted only to the limited tested scenarios and specific neurostimulation system used in these references [5, 13]. Caution is warranted in the generalization of this work and interpretation beyond the scope of these tested devices and scenarios.

The neurostimulation systems and the resultant heating appear to be sensitive to intrinsic as well as extrinsic factors. The configuration and/or geometry of the individual system (electrode, extension, and the IPG), the impedance of the wires, and wire breakage all can result in heating variances. Even, comparison of unilateral and bilateral systems may introduce a level of complexity. Extrinsic factors such as body versus head t/r coils, MR site landmarking, and SAR levels are also crucial considerations in the heating equation. Additionally, testing on externalized systems with externalized/percutaneous systems has revealed potential hazards of heating at higher SARs with extreme sensitivity to the routing of the extension cable. We are not aware of any published MR safety testing regarding these systems.

Other MR safety issues for neurostimulation systems regarding magnetic field interactions, induced voltages, programming changes, device functionality, and image artifacts need to be systematically addressed. Finally, it must be emphasized that a comprehensive checklist of precautions will be absolutely necessary to ensure safety for patients undergoing MR procedures.

A number of centers are performing MR procedures in patients with DBS systems under scenarios that have not been tested previously, while others are performing safety evaluations that are not yet comprehensive and rigorous and may be trivializing these safety issues. It is imperative not to forget the inherent risks posed to the patients and to undertake the responsibilities to fully explore these issues.

It is important that the neurosurgery, neuromodulation, and the MR healthcare communities to take a rigorous, critical, and standardized scientific approach towards MR safety with neurostimulators approach to the use of MR procedures in patients with neurostimulation systems and to help maintain the excellent safety record for the use of MR technology.

Acknowledgments

This investigation was funded by NIH/NINDS Grant NS044575-01.

References

1. Bligh J (1973) Temperature regulation in mammals and other vertebrates. American Elsevier, New York, pp 80–84
2. Chou C-K, Chan KW (1997) RF heating of implanted spinal fusion stimulator during magnetic resonance imaging. IEEE Transactions on Biomedical Engineering 44: 367–372
3. Cosman ER Radiofrequency lesions, in textbooks of stereotactic and functional neurosurgery. In: Gildenberg TR (ed) McGraw-Hill, New York, pp 973–985
4. Dormont D, Pidoux B, Bonnet AM, Biondi A, Oppenheim C, Hasboun D, Damier P *et al* (1997) Chronic thalamic stimulation with three-dimensional MR stereotactic guidance. Am J Neuroradiol 18: 1093–1097
5. Finelli D, Rugieri P, Tkach J, Nyenhuis JA, Hrdlicka G, Sharan A, Stypulkowski PH, Shellock FG (2002) MR-related heating of deep brain stimulation electrodes: an in vitro study using a head transmit/receive coil establishing the relationship between SAR and temperature elevation, with analysis of clinical imaging sequences. AJNR, Nov/Dec 23: 1–7
6. Gleason CA, Hricak H *et al* (1992) The effect of magnetic resonance imagers on implanted neurostimulators. Pacing Clin Electrophysiol 15: 81–94
7. Heistad DD (1974) Factors that influence blood flow in skeletal muscle and skin. Anesthesiology 41: 139–156
8. Houdas Y (1982) Temperature distribution, in Human Body

Temperature: its measurements and distribution. Plenum, New York
9. Kainz W, Uberbacher R, Alesch F, Chan DD (2002) Temperature Measurement on Neurological Pulse Generators during MR Scans. Bio Med Eng Online 1
10. Liem LA (1997) Magnetic resonance imaging and spinal cord stimulation systems. Pain 70: 95–97
11. Nutt JG, Peacock JH, Hammerstad JP, Burchiel KJ (2001) DBS and diathermy interactions induce severe CNS Damage. Neurology 56: 1384–1386
12. Ouayoun M, Aitbenamou C, Chouard CH (1997) Nuclear magnetic resonance and cochlear implant. Ann Otolaryngol Chir Cervicofac 114: 65–70
13. Rezai AR, Nyenhuis JA, Hrdlicka G, Tkach J, Sharan A, Rugieri P, Stypulkowoski PH, Shellock FG (2002) Neurostimulation systems for deep brain stimulation: in vitro evaluation of magnetic resonance imaging-related heating at 1.5 tesla. J Magn Reson Imaging 15: 241–250
14. Rezai AR, Crawley AP, Joy MLG, Davis KD, Kwan CL, Dostrovsky JO *et al* (1999) Thalamic stimulation and functional magnetic resonance imaging: localization of cortical and subcortical activation with implanted electrodes. J Neurosurg 90: 583–590
15. Rise MT (2000) Instrumentation for neuromodulation. Archi Med Res 31: 237–247
16. Shellock FG (2001) Magnetic resonance: health effects and safety. CRC Press Boca Raton, FL
17. Shellock FG (2001) MR imaging and electronically-activated devices. Radiology 219: 294–295
18. Shellock FG (2001) Pocket guide to metallic implants and MR procedures: update 2001. Lippincott-Raven Healthcare, New York
19. Soletra™ (Medtronic Inc, M, Physician and Hospital Staff Manual – Model 7426, http://www.medtronic.com/neuro/et/techmanual.html
20. Sommer T, Schimpf R *et al* (1998) MR in patients with cardiac pacemakers: in vitro and in vivo evaluation at 0.5 Tesla. Rofo Fortschr Geb Rontgenstr Neuen Bildgeb Verfahr 168: 36–43
21. Tronnier VM, Hahnel S, Sarem-Aslani A (1999) Magnetic resonance imaging with implanted neurostimulators: an in vitro and in vivo study. Neurosurgery 44: 118–125
22. Yamamoto T, Fukaya C *et al* (2000) Thalamotomy caused by cardioversion in a patient treated with deep brain stimulation. Stereotact Funct Neurosurg 74: 73–82
23. Zaremba L (2001) FDA Guidance for MR system safety and patient exposures: current status and future considerations, in magnetic resonance procedures: health effects and safety. Boca Raton, FL, CRC Press, pp 183–196
24. Zonenshayn M, Rezai AR (2000) Neurostimulation and functional brain stimulation. Neurol Res 22: 318–325

Correspondence: Ali R. Rezai, M.D., Department of Neurosurgery S-80, The Cleveland Clinic Foundation Cleveland, Ohio 44195. E-mail: rezaia@ccf.org

Pain control

Acta Neurochir (2003) [Suppl] 87: 149–152

Primary motor cortex stimulation within the central sulcus for treating deafferentation pain

Y. Saitoh[1], **A. Kato**[1], **H. Ninomiya**[1], **T. Baba**[1], **M. Shibata**[2], **T. Mashimo**[2], and **T. Yoshimine**[2]

[1] Department of Neurosurgery, Osaka University Medical School, Osaka, Japan
[2] Department of Anesthesiology, Osaka University Medical School, Osaka, Japan

Summary

Nine patients with post-stroke pain, six with brachial plexus injuries, two with phantom limb pain, one with spinal cord injury, and one with brain stem injury were treated with a modified motor cortex stimulation (MCS) protocol. Preoperative pharmacological tests were performed with phentolamine, lidocaine, ketamine, thiopental, morphine, and placebo. We placed a grid electrode in the subdural space to decide upon the best stimulation point for pain relief over a few weeks with the purpose of determining the placement of a Resume electrode. In five patients, Resumes were implanted in the interhemispheric fissure to reduce lower extremity pain. In five other patients, Resumes were placed within the central sulcus to stimulate area 4 and area 3b. In addition, electrodes were also placed on the surface of the precentral gyrus. Fourteen of the 19 patients showed pain reduction (6 excellent, 3 good, and 5 fair) using the MCS with our results indicating area 4 within the central sulcus to be the optimal stimulation point for pain relief. We speculate that conventional method may sometimes fail to stimulate area 4 and that focal stimulation of the primary motor cortex within the central sulcus may improve the efficacy of this treatment. Our pharmacological tests show that patients with ketamine sensitivity seem to be good candidates for MCS. Test stimulation with a subdural multi-grid electrode and Resumes in the cetral sulcus were helpful in locating the best stimulation point for pain relief.

Keywords: Thalamic pain; phantom limb pain; brachial plexus injury; motor cortex stimulation.

Introduction

Deafferentation pains have been the most difficult type of pain to treat. Katayama *et al.* reported that the use of motor cortex stimulation (MCS) may provide relief from pain in approximately 50% of patients with thalamic pain [8]. On the other hand, Meyerson *et al.* showed that facial deafferentation pain may be decreased using MCS [10]. We have also previously reported MCS to be effective for treating peripheral deafferentation pain (brachial plexus injuries and phantom limb pain) [16].

We used a protocol in which, under general anesthesia, a grid electrode or 4-array electrodes (Resume, Medtronic) were placed in the subdural space for test stimulation. Various stimulation patterns were tested over a period of a few weeks, after which patients experiencing pain relief underwent implantation of a Resume at the position determined as the most effective. We implanted Resumes into five patients to stimulate areas 4 and 3b within the central sulcus.

A prior report on preoperative pharmacological tests (DCTs) concerning the relief of pain has suggested that thiopental- and ketamine-responsive, as well as morphine-resistant patients receive lasting benefits from MCS [23]. We have examined the correlations between the results of DCTs and the efficacy of stimulation using a somewhat different protocol to that of the above [17].

Patients and methods

Patient population

Nineteen patients with central and peripheral deafferentation pain were treated with MCS at Osaka University Hospital and Osaka Rosai Hospital between January 1996 and May 2002. Patients included 16 men and three women whose ages ranged from 30 to 68 years. Among these patients, seven had histories of thalamic hemorrhage, one of putaminal hemorrhage, one of pontine hemorrhage, six of brachial plexus injury, two of phantom limb pain, one of spinal cord injury, and one of pontine injury. The interval between hemorrhaging and the onset of pain ranged from 0.5 to 4 years. Patients were treated with various medications including NSAIDs, anticonvulsants and antidepressants. These drugs failed to sufficiently reduce pain despite their use in various combinations.

All the patients with central pain also suffered from varying degrees of hemiparesis. The patients complained of intense spontaneous pain in the extremities or trunk with a burning or tearing sensation. Various degrees of pinprick hypesthesia were demonstrable in all patients except those with phantom limb pain. No patients psychologically assessed to be showing severely depressive or neurotic tendencies were included in the study.

Pharmacologic tests (Drug challenge tests; DCTs) and pain assessment

In an attempt to clarify the pathophysiologic mechanisms in these patients, DCTs were performed. Their detailed protocols are as previously described [17].

More than 20 different stimulatory patterns were evaluated with use of the grid electrodes or Resumes to determine the best stimulation point for pain relief. Each patient was asked to describe pain intensity according to a visual analogue scale (VAS) and the McGill Pain Questionnaire. The effects of stimulation were classified into four categories according to the percentage of reduction in pain from prestimulation levels: excellent, 80 to 100% reduction; good, 60 to 79% reduction; fair, 40 to 59% reduction; and poor, <40% reduction. The pain level was evaluated after no more than 30 minutes stimulation.

Surgical procedures

The location of the central sulcus was approximated by a preoperative MRI. Under general anesthesia, craniotomy was performed overlying the motor and sensory cortex. In 16 patients, a 20-grid electrode (4 × 5 array; 0.3 cm electrode diameter; 0.7 cm separation; Unique Medical Co., Tokyo, Japan) was placed subdurally to confirm the locations of the central sulcus by the observation of the phase reversal of the N20 component with median nerve stimulation. Five patients received Resumes (0.4 cm diameter, 0.6 cm separation; Medtronic, Inc., Minneapolis, MN) in the central sulcus to stimulate the areas 4 and 3b after microscopic dissection of the central sulcus. The five patients with leg pain underwent placement of a Resume in the interhemispheric fissure.

After implantation of the electrodes, electrical stimuli consisting of monophasic square-wave pulses 0.2 msec in duration were delivered to various areas. The frequency (25 to 50 Hz) and voltage (0.9 to 5 V) producing the best pain relief were determined. In the case of the implantation of Itrel III, impedance was between 900 and 1500 ohms. Stimulation was usually applied continuously for 30 minutes on each occasion. If a patient developed seizures upon stimulation, the stimulatory voltage was decreased to a level lower than the threshold for muscle contraction. After definitive electrode implantation, chronic stimulation was delivered with a wireless stimulation system (X-trel model 3425, Máttrix model 3210, Itrel III IPG model 7425, Medtronic Inc., Minneapolis, MN).

Results

Fourteen of the 19 patients showed various degrees of pain control in test stimulations. In the other five, more than 20 different patterns of stimulation on the electrodes failed to reduce pain. All five patients, who received Resumes within the central sulcus, preferred stimulation of area 4 within the central sulcus over those of area 3 and the surface of precentral gyrus. When electrode placement was in the interhemispheric fissure, possible electrode positions were limited. In each patient with successful responses, several stimulation points were effective, and pain relief was achieved even with stimulation of the sensory cortex. However, stimulation of the premotor cortex failed to relieve pain. In two patients complaining of pain extending from the upper extremity to the hip, and phantom foot pain and stump pain, it was found to be difficult to provide complete treatment with a single electrode. In these cases, Máttrix dual-electrode systems were found to bring relief from pain after implantation following confirmation of the most effective stimulation points.

Of the 18 final participants in the DCTs, eight cases scoring with "excellent" or "good" pain relief using the MCS were found to have sensitivities to morphine (n = 5), ketamine (n = 4), thiopental (n = 4) or lidocaine (n = 3). The other 10 cases scoring "fair" or "poor" pain relief had morphine (n = 4) or thiopental (n = 2) sensitivities. No relationship was found between morphine sensitivity and pain relief with the MCS, and none of the patients was found to be sensitive to phentolamine. Two of the five patients with excellent pain relief did not respond to any drug. From these results, we conclude that ketamine may be a useful drug for anticipating the effectiveness of the MCS in the treatment of deafferentation pain.

Three patients showed pain relief for 24 hours with 30 minutes' stimulation. In contrast, two patients achieved relief for only about 1 hour after electrical stimulation. The others obtained pain relief for 3–5 hours. Patients were observed as outpatients for periods of 6–50 months. Two patients suffered cerebral hemorrhage; one is now in a vegetative state, and the other died 36 months after Resume implantation. The whole systems were removed from 2 patients because they suffered from infections of the stimulation systems about 6 months after implantation.

Discussion

Previous reports have described implantation of an epidural Resume over the precentral gyrus [1, 2, 4, 8, 11, 14, 21]. Such an approach might not provide optimal pain relief since both the method and the area of test stimulation were restricted by a brief operative period under local anesthesia. We chose a two-stage procedure in which we typically used a 20-grid electrode over the sensory-motor cortex to test various geometric patterns of stimulation with the purpose of determining the optimal pattern for pain relief. In five patients, Resumes were implanted in the central sulcus in addition to the surface of the precentral gyrus. The second stage of our procedure involved the implantation of a Resume based on the results of the first stage.

The optimal placement and orientation of a Resume in relation to the motor cortex has not yet been established [12]. In our test stimulations, stimulation of the prefrontal cortex failed to reduce pain in all cases, whereas stimulation of the sensory cortex did reduce pain somewhat. In all effective cases, the optimal stimulation site for pain relief was on the surface of the precentral gyrus and area 4 within the central sulcus. Stimulation of area 4 within the central sulcus seemed to be more effective than those of area 3b and the surface of precentral gyrus. In our series, the duration of pain relief with 30 minutes' stimulation ranged from 1 to 24 hours. This wide variation in the period of pain relief may be explained by several factors, including the results of DCTs, the original pathologies and the stimulation point.

VAS was used to evaluate the stimulation effect in our study. In test stimulations, patients were repeatedly and randomly asked to score the effect according to the VAS at stimulation, and these results were fairly consistent. Patients described whether pain was reduced or worsened in comparison to that prior to stimulation and rated the intensity and nature of the pain in accordance with the McGill Pain Questionnaire before and after stimulation. It appeared that the VAS changes for each patient were reliable. Testing with a grid electrode or Resumes can control psychological effects.

On the Brodmann brain map, the parts of areas 4 and 3b corresponding to the hand are very narrow at the cortical surface. The major parts are located within the central sulcus (Fig. 1). There are reportedly strong connections between area 4 and the sensory areas (areas 2, 3a, 5 and the secondary sensory area). However, there are few fiber connections between areas 4 and 3b [20]. In this study, focal stimulation of area 3b within the central sulcus was confirmed to be less effective than that of area 4. Functional mapping has recently shown that the motor and sensory cortices overlap and are not divided in a simple manner by the central sulcus [13]. Therefore, conventional stimulation of the cortical surface might stimulate cortical areas other than area 4, and some patients have shown greater pain reduction in response to stimulation of area 4 in addition to the effect achieved by stimulation of other areas including the sensory cortex. In the present study, stimulation of the area 4 within the

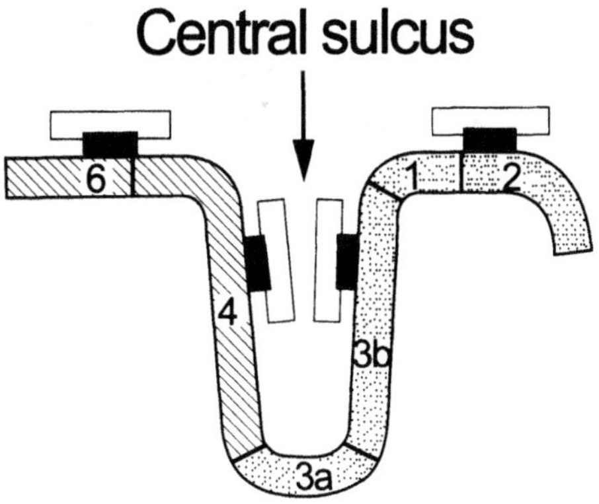

Fig. 1. The major part of area 4 (primary motor cortex) is located within the central sulcus, and the surface portion may be too small to influence the upper extremities and face. In this study, focal stimulation of area 3 and 4 was performed

central sulcus was considered the optimal site for reducing deafferentation pain.

From our DCTs, ketamine was found to be a possible drug useful for anticipating the effectiveness of MCS for treating deafferentation pain. However, two of the five patients with excellent results did not respond to any drug. It should be mentioned that our DCT is different from those of previous protocols [9, 23]. It is thought that a uniform method of DCT is needed.

Our present results suggest that MCS may be effective for both peripheral and central deafferentation pain. Although MCS has been reported effective for facial deafferentation pain [10, 12], it has been rarely studied in cases of phantom-limb pain or brachial plexus avulsion [17]. Phantom-limb pain is a very rare symptom for which several medical [3, 7, 18, 19] and surgical [15] approaches have been pursued. Postcentral stimulation of the appropriate areas can actually exacerbate phantom-limb pain in some patients. Even with excision of cortical or thalamic areas, phantom-limb pain tends to recur with time [5]. DREZotomy (dorsal root entry zone lesion) which is a more invasive surgery than MCS is reported effective in 60 to 70% of brachial plexus avulsions [4]. Notably, there was one case that did not respond to DREZotomy but responded well to stimulation.

MCS has been found effective in approximately half of patients with central deafferentation pain in the largest published study [8] together with many other reports [1, 2, 11, 14]. Our success rate is better than those of others, partly because we attempted stimulating area 4 within the central sulcus. We plan to continue using and evaluating stimulation in the central sulcus. This strategy may improve the success rate for treatment of patients with deafferentation pain.

Acknowledgment

We greatly appreciate the help of Dr. Shayne Morris for his suggestions and editing of our manuscript.

References

1. Canavero S, Bonicalzi V (1995) Cortical stimulation for central pain. J Neurosurg 83: 1117
2. Ebel H, Rust D, Tronnier V, Boker D, Kunze S (1996) Chronic precentral stimulation in trigeminal neuropathic pain. Acta Neurochir (Wien) 138: 1300–1306
3. Elliott F, Little A, Milbrandt W (1976) Carbamazepine for phantom limb phenomena. N Engl J Med 295: 678
4. Gybels JM (1992) Indications for Neurosurgical treatment of chronic pain. Acta Neurochir 116: 171–175
5. Horrax G (1946) Experiences with cortical excisions for the relief of intractable pain. Surgery 20: 593–602
6. Hosobuchi Y (1986) Subcortical electrical stimulation for control of intractable pain in humans. Report of 122 cases (1970–1984). J Neurosurg 64: 545–553
7. Jaeger H, Maier C (1992) Calcitonin in phantom limb pain: a double-blind study. Pain 48: 21–27
8. Katayama Y, Fukaya C, Yamamoto T (1998) Poststroke pain control by chronic motor cortex stimulation: neurologic characteristics predicting a favorable response. J Neurosurg 89: 585–591
9. Meyerson BA (1997) Pharmacological tests in pain analysis and in prediction of treatment outcome. Pain 72: 1–3
10. Meyerson BA, Lindblom U, Linderoth B, Lind G, Herregodts P (1993) Motor cortex stimulation as treatment of trigeminal neuropathic pain. Acta Neurochir (Wien) [Suppl] 58: 150–153
11. Migita K, Uozumi T, Arita K, Monden S (1995) Transcranial magnetic coil stimulation of motor cortex in patients with central pain. Neurosurgery 36: 1037–1040
12. Nguyen JP, Keravel Y, Feve A, Uchiyama T, Cesaro P, Le Guerinel C, Pollin B (1997) Treatment of deafferentation pain by chronic stimulation of the motor cortex: report of a series of 20 cases. Acta Neurochir (Wien) [Suppl] 68: 54–60
13. Nii Y, Uematsu S, Lesser RP, Gordon B (1996) Does the central sulcus divide motor and sensory functions? Cortical mapping of human hand areas as revealed by electrical stimulation through subdural grid electrodes. Neurology 46: 360–367
14. Peyron R, Garcia-Larrea L, Deiber MP, Convers P, Sindou M, Mauguiere F, Laurent B (1995) Electrical stimulation of precentral cortical area in the treatment of central pain: electrophysiological and PET study. Pain 62: 275–286
15. Raj PP (1992) Practical management of pain, 2nd ed. Year Book Medical Publisher, St Louis
16. Saitoh Y, Shibata M, Sanada H, Mashimo T (1999) Motor cortex stimulation for phantom limb pain. Lancet 353: 212
17. Saitoh Y, Shibata M, Hirano S, Hirata M, Mashimo T, Yoshimine T (2000) Motor cortex stimulation for central and peripheral deafferentation pain. Report of eight cases. J Neurosurg 92: 150–155
18. Smith PP, Turkington D (1993) Fluoxetine in phantom limb pain. Br J Psych 163: 105–106
19. Stannard CF, Porter GE (1993) Ketamine hydrochloride in the treatment of phantom limb pain. Pain 54: 227–230
20. Stepniewska I, Preuss TM, Kaas JH (1993) Architectonics, somatotopic organization, and ipsilateral cortical connections of the primary motor area (M1) of owl monkeys. J Comp Neurol 330: 238–271
21. Tsubokawa T, Katayama Y, Yamamoto T, Hirayama T, Koyama S (1993) Chronic motor cortex stimulation in patients with thalamic pain. J Neurosurg 78: 393–401
22. Wood CC, Spencer DD, Allison T, McCarthy G, Williamson PD, Goff WR (1988) Localization of human sensorimotor cortex during surgery by cortical surface recording of somatosensory evoked potentials. J Neurosurg 68: 99–111
23. Yamamoto T, Katayama Y, Hirayama T, Tsubokawa T (1997) Pharmacological classification of central post-stroke pain: comparison with the results of chronic motor cortex stimulation therapy. Pain 72: 5–12

Correspondence: Youichi Saitoh, M.D., Department of Neurosurgery, Osaka University Medical School, 2-2 Yamadaoka, Suita, Osaka 565-0871, Japan. E-mail: saitoh@nsurg.med.osaka-u.ac.jp

Acta Neurochir (2003) [Suppl] 87: 153–155

Neurons with spontaneous high-frequency discharges in the central nervous system and chronic pain

K. Yamashiro[1], **N. Tomiyama**[2], **Y. Terada**[3], **H. Samura**[4], **J. Mukawa**[1], and **R. R. Tasker**[5]

[1] Department of Neurosurgery, University of the Ryukyus School of Medicine, Okinawa Japan
[2] Department of Neurosurgery, Okinawa Prefectual Naha Hospital, Okimawa, Japan
[3] Department of Neurosurgery, Ohhama Daiichi Hospital, Okinawa, Japan
[4] Department of Neurosurgery, Okinawa Central Hospital, Okinawa, Japan
[5] Division of Neurosurgery, Toronto Hospital University of Toronto, Toronto, Canada

Summary

Neurons with high-frequency discharge (hyperactive neuron) were recorded in the thalamus of chronic pain patients. We created a chronic pain rat model, and recorded the discharge of high-frequency neurons by using a microelectrode in the thalamic area. The character of these high-frequency discharge neurons was assessed.

Deep brain stimulation (DBS) was done in 9 patients, and motor cortical stimulation (MCS) in 2. Prior to implantation of the stimulation electrodes, extracellular unitary activity was recorded and local microstimulation was done by the same microelectrode. From the ventrocaudal (Vc) nucleus of the chronic pain patients, hyperactive neurons were recorded. There were three types of discharge patterns in interval histograms. Some hyperactive neurons showed firing suppression for a short time period by intravenous administration of phenytoin and calcium antagonist (Nicardipine).

In animal experiments, unilateral dorsal root sectioning from C5 to Th1 were made in male Wistar rats according to the method of Lombard *et al.* (1979). A few months after the operation, hyperactive neurons were recorded from the thalamic nuclei and lemniscus medialis. The firing patterns and distribution of hyperactive neurons were very similar to those of humans. Sensorimotor cortical electrical stimulation showed a reduction of firing in the hyperactive neurons. Iontophoretical application of glutamate yielded an increase in firing. In contrast, GABA and NMDA antagonist MK-801 revealed remarkable firing suppression.

These results suggest that hyperactive neurons may correlate with the glutamatergic, especially NMDA and GABAergic receptor or fibers.

Keywords: Hyperactive neuron; pain; thalamus; neurotransmitter.

Introduction

Following injury, stimulation of peripheral nerves, spinal cord and brain sometimes elicits hyperalgesia and/or allodynia in animals and humans. In this state, spontaneous multiple firing (hyperactive) neurons have been recorded within the posterior horn of the spinal cord, brain stem, thalamus, and cortex in animal experiments [1, 2, 8, 10], but also in the posterior horn of the spinal cord and in the thalamic nuclei in patients with chronic pain [7, 9, 12, 14]. These phenomena suggest that there may be characteristic changes in the central nervous system (CNS) [1, 2, 7–10, 12–14]. To investigate such possible changes, we examined the extra-cellular responses. The goal of this study was to examine hyperactive neurons with high frequency discharge in the human brain during surgery and animal thalamic nuclei cell responses by chemical substances and electrical stimulation.

Materials and methods

Eleven patients with chronic pain were treated by deep brain stimulation (DBS). In five patients pain resulted from cerebrovascular disease, in two from spinal cord injury, and in one each from peripheral nerve injury, amputation (stump pain), trigeminal neuralgia and multiple sclerosis. Extracellular unitary recording and local microstimulation (100 Hz, 0.2 ms, up to 120 uA) were performed in physiological localization of the DBS electrode in the thalamic nuclei in the target area, using the same microelectrode [14, 18]. Responses in general were correlated with those surrounding any hyperactive neurons found mainly ventrocaudally. After completion of localization, a DBS electrode was inserted (Medtronics Corp.).

In the animal experiments, unilateral dorsal root sectioning from C5 to Th1 in male Wistar rats was performed according to the methods and chronic pain model of Lombard *et al.* [10]. One to three months after the dorsal root sectioning, the animal showed self-mutilation of the right forepaw or shoulder area. During this period, electrophysiological experiments were performed. Cellular behaviors were recorded from the contralateral thalamic nuclei, lemniscus medialis and internal capsule and also some kind of neurotransmitter effect was examined.

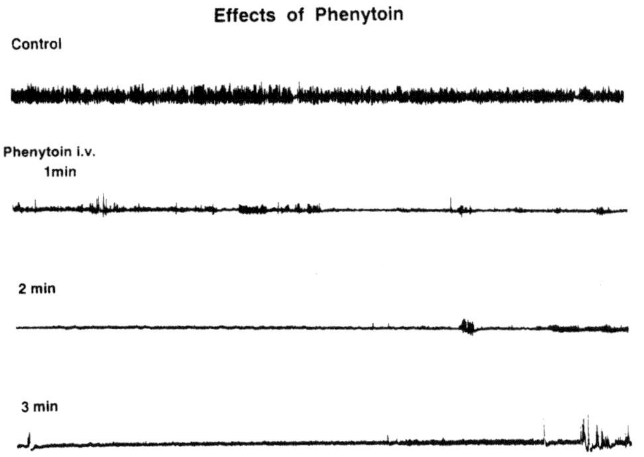

Fig. 1. Effect of phenytoin on the human hyperactive neuron. After intravenous administration of phenytoin, firing of the hyperactive neuron was suppressed for a few minutes

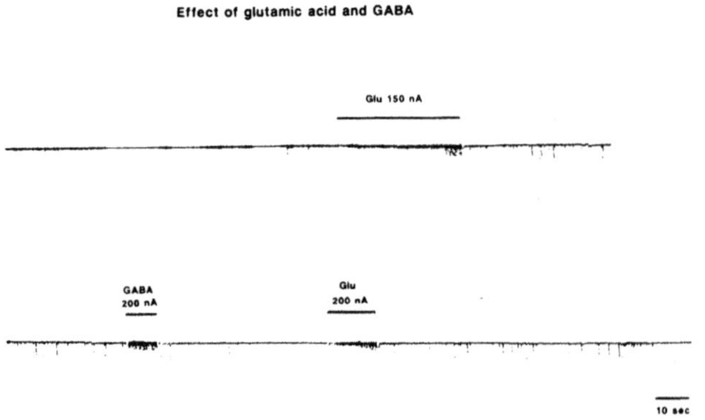

Fig. 2. Effect of iontophoretical glutamate and GABA application on the hyperactive neuron in the thalamic nucleus of denervated rat. Glutamate increases firing rate. GABA suppresses the firing rate markedly

Results

From the human thalamus, three types of hyperactive neurons (which have different firing patterns) were recorded. Type A has slow continuous firing with 100 msec peak in the interval histogram. Type B shows a more frequent firing rate as compared to A. Type C neuron has two peaks, 5 msec and about 200 msec, and this type showed regular grouping discharge. These hyperactive neurons were distributed in ventrocaudal (Vc), ventral intermediate (Vim), ventral oral posterior (Vop) nuclei of thalamus. Usually these neurons showed little effect by peripheral stimulation. While administration of phenytoin and Ca antagonist, Nicardipine showed a remarkable decrease of firing of hyperactive neurons (Fig. 1).

In the animal experiments, spontaneous firing hyperactive neurons were recorded from the contralateral VP nucleus of the rat. These spontaneously firing hyperactive neurons have two peaks in the interval histogram analysis, which is 2–3 msec in the short interval histogram, and 200 to 400 msec peak in the long interval histogram. The firing patterns of rats were very similar to those of humans. These hyperactive neurons showed a decrease of firing by electrical stimulation to the sensorimotor cortex. Most suppression was gained at 50 and 100 Hz stimulation rate. Diazepam and phenytoin were administered intravenously. Both drugs showed suppression of firing. Iontophoretical application of GABA, glutamate and NMDA antagonist MK-801 was performed to the hyperactive neurons. Application of glutamate to the hyperactive neurons showed an increase of firing, but the application of GABA to the hyperactive neuron yielded a remarkable reduction of firing (Fig. 2). Application of MK-801 showed a remarkable suppression of firing.

Discussion

High frequency discharge neurons in the thalamic nuclei and in the spinal posterior horn in patients with chronic pain have been reported [8, 10, 13, 15].

In animal experiments, it has also been reported that after injury to peripheral sensory nerves, high frequency discharge neurons were observed at the posterior horn, brain stem, thalamus and cerebral cortex [1, 2, 8, 10]. Loeser outlined a correlation between high frequency discharge neuron and pain [9].

Gorecki *et al.* [8] recorded bursting cells from the thalamic nuclei of chronic pain patients. They were classified into three types based on their firing pattern. Hyperactive neurons were distributed at the inferocaudal of ventral caudal (Vc) and rostro-medial and caudo-lateral of ventral intermediate (Vim) nuclei of the thalamus in our chronic pain patients. They were 15% in frequency of our recording cells and classified into three types of firing pattern as Gorecki *et al.* [7, 18].

Many neurotransmitters and chemical substances were reported as being in correlation with pain. Excitatory amino acids, such as glutamate and GABA were also reported to correlate with pain [5, 13, 15, 18]. It was reported that GABA has anti-nociceptive action [11]. GABA binds three distinct receptor subtypes:

GABAV, GABAV, and GABAc. Because they open bicuculline-sensitive Cl-channels, GABAV receptors are responsible for most of the fast inhibitory synaptic transmission in the brain. Structurally, GABAV receptors are presumably heteropentameric structures [16]. In our experiment, the activity of hyperactive neurons in the rat thalamus was suppressed by GABA, phenytoin, and diazepam. These results suggest that the hyperactivity of neurons may correlate with GABAergic activity and that the GABAV receptor plays an important role in the suppression of hyperactivity of neurons. Glutamate subtype, NMDA receptor correlate to the perception of pain and wind-up phenomenon [15].

Non-competitive (MK-801) and competitive NMDA receptors antagonists prevented the wind-up phenomenon and autotomy [13, 17]. These reports indicate that NMDA receptors have a potential role in the induction and maintenance of chronic pain. Bordi and Quartaroli [4] suggest that block of NMDA receptor decreases nociceptive transmission in the thalamus and can modulate hyperalgesic states. Glycine antagonist also blocked responses to noxious stimuli. Our observations of hyperactive neuron responsiveness to glutamate, NMDA antagonist and GABA at the thalamus were supported in as much as that very similar processes may occur at the thalamic hyperactive neurons. Some hyperactive neurons had a decreased firing after Phenytoin, calcium antagonist Nicardipine administration. It was reported that Phenytoin accerelates the sodium efflux from the cell and correlates with chloric ion channel. It was reported that Nicardipine is a calcium entry blocker (slow channel blocker) that inhibits the transmembrane influx of calcium ions [6]. These results suggest that hyperactive neurons may correlate with the glutamatergic, especially NMDA and GABAergic receptor or fibers.

References

1. Albe-Fessard D, Lombard MD (1983) Use of an animal model to the origin of and projection against deafferentation pain. In: Bonica JJ, Lindblom U, Iggo U (eds) Advances in pain research and therapy, vol 5. Raven Press, New York, pp 691–700
2. Anderson LS, Black RG, Abraham J, Ward AA Jr (1972) Neuronal hyperactivity in experimental trigeminal deafferentation. J Neurosurg 35: 444–452
3. Behbehani MM, Jiang M, Chandler SD, Ennis M (1990) The effect of GABA and its antagonists on midbrain periaqueductal gray neurons in the rat. Pain 40: 195–204
4. Bordi F, Quartaroli M (2000) Modulation of nociceptive transmission by NMDA/glycine site receptor in the ventropostero-lateral nucleus of the thalamus. Pain 84: 213–224
5. Coderre TJ, Katz J, Vaccarino AL, Melzack R (1993) Contribution of central neuroplasticity to pathological pain: review of clinical and experimental evidence. Pain 52: 259–285
6. Goodfraind T, Miller R, Wibo M (1986) Calcium antagonism and calcium entry blockade. Pharmacol Rev 38: 321–416
7. Gorecki J, Hirayama T, Dostrovsky JO, Tasker RR, Lenz FA (1989) Thalamic stimulation and recording in patients with deafferentation and central pain. Stereotact Funct Neurosurg 52: 219–226
8. Loeser JD, Ward AA Jr (1967) Some effects of deafferentation neurons of the cat spinal cord. Arch Neurol 17: 629–636
9. Loeser JD, Ward AA Jr, White LE Jr (1968) Chronic deafferentation of human spinal cord neurons. J Neurosurg 29: 48–50
10. Lombarad MC, Nashold BS Jr, Albe-Fessard D (1979) Deafferentation hypersensitivity in the rat after dorsal rhizotomy: a possible animal model of chronic pain. Pain 6: 163–174
11. Rechling DB, Basbaum AI (1990) Contribution of brainstem GABAergic circuitry to descending antinociceptive controls: II. Electron microscopic immunocyto-chemical evidence of GABAergic control over the projection from the periaqueductal gray to the nucleus raphe magnus in the rat. J Comp Neurol 302: 378–393
12. Rinaldi PC, Young RF, Albe-Fessard D, Chodakiewitz J (1991) Spontaneous neuronal hyperactivity in the medial and intralaminar thalamic nuclei of patients with deafferentation pain. J Neurosurg 74: 415–421
13. Seltzer Z, Cohn S, Ginzburg R, Beilin B (1991) Modulation of neuropathic pain behavior in rats by spinal disinhibition and NMDA receptor blockade of injury discharge. Pain 45: 69–75
14. Tasker RR, Lenz FA, Yamashiro K, Gorecki J, Hirayama T, Dostrovsky JO (1987) Microelectrode techniques in localization of stereotactic targets. Neurol Res 9: 105–112
15. Thompson SWN, Woolf CJ (1991) Primary afferent-evoked prolonged potentials in the spinal cord and their central summation: role of the NMDA receptor. In: Bond MR, Charlton JE, Woolf CJ (eds) Proceedings of VIth World Congress on pain. Elsevier, New York, pp 291–297
16. Wan Q, Man H Ye, Braunton J, Wang W, Salter MW, Becker L, Wang YT (1997) Modulation of GABAA receptor function by tyrosine phosphorylation of β subunits. J Neurosci 17: 5062–5069
17. Woolf CJ, Thompson SWN (1991) The induction and maintenance of central sensitization is dependent on N-methyl-D-aspartic acid receptor activation; implications for the treatment of post-injury pain hypersensitivity states. Pain 44: 293–299
18. Yamashiro K, Mukawa J, Terada Y, Tomiyama N, Ishida A, Mori K, Tasker RR, Albe-Fessard D (1994) Neurons with high-frequency discharges in the central nervous system in chronic pain. Stereotact Func Neurosurg 62: 290–294

Correspondence: Katsumi Yamashiro, M.D., Department of Neurosurgery, University of the Ryukyus Faculty of Medicine, 207 Uehara, Nishihara, Okinawa 903-0125, Japan. E-mail: yamaka@med.u-ryukyu.ac.jp

Nerve grafting and cell transplantation

Acta Neurochir (2003) [Suppl] 87: 159–162

First human ventral mesencephalon and striatum cografting in a parkinson patient

C. C. Lee[1], S. Z. Lin[1], Y. Wang[2], J. J. Lin[3], J. Y. Liu[4], G. J. Chen[5], Y. H. Chiang[5], J. C. Liu[6], and F. C. Zhou[7]

[1] Department of Radiology, Neuro-Medical Scientific Center, Tzu-Chi General Hospital, Tzu-Chi University, Hualien, Taiwan
[2] National Institute on Drug Abuse, NIH, USA
[3] Department of Neurology, Chushang Show-Chwan Hospital, Nantou, Taiwan
[4] Department of Gynecology, Tri-Service General Hospital and National Defense Medical Center, Taipei, Taiwan
[5] Department of Neurological Surgery, Tri-Service General Hospital and National Defense Medical Center, Taipei, Taiwan
[6] Department of Anatomy, Tri-Service General Hospital and National Defense Medical Center, Taipei, Taiwan
[7] Department of Anatomy, Indiana University School of Medicine, Indianapolis, USA

Summary

Fetal ventral mesencephalon (VM) transplantation has been reported to improve Parkinsonian symptoms. Animal studies show that cografting of striatal tissue increases the survival of dopamine neurons. Whether or not VM and striatum cografting could ameliorate motor dysfunction in a Parkinson's disease (PD) patient was explored in this study. The patient was a 53-year-old male who had presented with symptoms of tremor, rigidity and bradykinesia for 11 years. He had been treated with L-dopa and had progressive deterioration of symptoms even with the daily dosage of L-dopa increased to 900 mg per day. Before transplantation, his PD symptoms were scored with Unified Parkinson's Disease Rating Scale (UPDRS) and video recordings. The influx constant (ki) of the [18F] 6-fluoro-L-dopa uptake in the striatum was measured by positron emission tomography (PET) imaging. The fetal VM and the lateral part of the lateral ganglionic eminence (LGE) were cografted into the right putamen and, one week later, fetal VM alone was transplanted into the left putamen. After the transplantation, the patient's UPDRS score improved from 128 to 62 at 6 months and to 24 at 22 months during the "off" phase. The score of daily living disability improved from 35 to 18 at 6 months and to 10 at 22 months post transplantation. Twenty-two months after grafting, "off" phases were almost absent, and the freezing had totally disappeared. The [18F] 6-fluoro-L-dopa PET studies were performed 1 month before and 21 months after transplantation. The ki for [18F] 6-fluoro-L-dopa was decreased by 15% in the right caudate and 5% in the left caudate, both of which did not have any ventral mesencephalic grafts. However, the ki was increased by 35% in the left non-cografted putamen, and by 58% in the right cografted putamen. In conclusion, cografting the fetal VM and the LGE in the putamen may improve the motor function of PD patients.

Keywords: Parkinson's disease; transplantation; ventral mesencephalon; striatum; cograft.

Introduction

PD is a progressive neurodegenerative disease presented with symptoms of akinesia, rigidity, tremor and postural instability [1, 5, 9, 13, 14]. During the early stage of PD, L-dopa is effective for symptomatic control. However, as the disease progresses, the symptoms of PD become poorly controlled with L-dopa [4, 7]. To improve the treatment efficacy for these PD patients, fetal VM transplantation has been performed in a number of them [1, 4, 9, 14]. Variable but significant symptomatic improvement after the neural grafting has been reported in several studies [1, 4, 9, 14]. In addition, histological evidence of long term survival of the transplanted fetal VM has been identified in a postmortem study [3].

In order to achieve effective symptomatic improvement, 3 to 4 fetal VMs were usually needed for each side of the putamen [3]; therefore, a total of 7 fetal VMs were required for one PD patient [3]. As the survival rate of the grafted neurons is only about 10%, this large amount of tissue is needed for each patient to ensure an improvement in PD symptoms [3, 6]. Most dopamine neurons die within 24 hours after grafting in the putamen [3, 6].

Because of the limited supply of human fetal VM, there have been many studies focussing on enhancing the survival rate of dopamine neurons. Cografting dopamine neurons with striatal tissue increases not only the survival rate of dopamine neurons but also the density of outgrowing axons from the grafted dopamine neuron into the striatal tissue [10, 15]. The cografting effect may be related to the secretion of trophic factors from the fetal striatal tissue. These factors might include glial cell line-derived neurotrophic factor (GDNF) and brain derived neurotrophic factor

(BDNF) [8, 10, 15]. Based on the above findings, we examined whether cografting fetal VM with fetal striatal tissue might increase the efficacy of grafted dopamine neurons in PD patients.

Case report

The patient was a 53-year-old man presenting with symptoms of rigidity, tremor and akinesia for 11 years. Initially, these symptoms began in the left limbs and progressed to both sides within two years. At that time, he was diagnosed as having PD and started L-dopa therapy. Due to worsening of the symptoms, the patient gave up his job 8 years ago. Postural instability and motor fluctuations continued to worsen thereafter. One year before his admission, he went to a hospital in Japan for a left pallidotomy. However, only a transient (7 days) improvement of symptoms was obtained. At the present admission, the patient was graded as stage 4 on the Hoehn and Yahr Scale. He was treated with L-dopa 100 mg and carbidopa 25 mg five times a day, long acting L-dopa 100 mg and benserazide 25 mg four times a day. The patient still had marked symptoms of postural instability and freezing phenomenon. He was in an "off" phase 60 percent of the time and the UPDRS score was 128 during "off" phases. The Activity of Daily Living of UPDRS was 35 during "off" and 12 during "on" phases. A magnetic resonance imaging of the brain revealed no significant findings except a tiny ovoid lesion, about 2 mm in diameter, in the posterior part of the left globus pallidus due to the previous pallidotomy. A [18F] 6-fluoro-L-dopa PET study, done at the PET center of the University of British Columbia in Vancouver, Canada, demonstrated a typical picture of PD [3, 11]. The uptake of [18F] 6-fluoro-L-dopa was markedly decreased bilaterally in the putamen. The ki of [18F] 6-fluoro-L-dopa in the right and left caudates were 0.0060 and 0.0063 (normal 0.0093 ± 0.0010), respectively, and in the right and left putamen 0.0021 and 0.0027 (normal 0.0079 ± 0.0009), respectively.

Under local anesthesia, neural transplantation was performed bilaterally in two operations 7 days apart. A burr hole, about 6 mm in diameter, was made in the frontal region, 2.5 cm away from the midline and 12.0 cm behind the nasion. The VM and the right LGE were dissected from a fetal brain aged 7.5 weeks. For cografting, the LGE was cut into many small pieces ($0.1 \times 0.1 \times 0.1$ mm^3). The VM was cut into three blocks. Each block, together with few tiny pieces of the LGE, was transplanted into the right post commissural putamen from the ventral to the dorsal part. The post commissural putamen was evenly divided into three regions for cografting the three blocks of the fetal VM. A Leksell stereotactic device (G frame) with magnetic resonance imaging localization was used for grafting. The needle for transplanting fetal tissue was 1 mm in diameter, and a Hamilton syringe of 250 μl was used for injecting the fetal tissue. Immediately after the neural transplantation, follow-up magnetic resonance imaging was performed to ensure the proper localization. One week before the neural transplantation, the patient was treated with cyclosporin-A (6 mg/kg/day). Seven days after the right side transplantation, another fetal ventral mesencephalic transplantation was performed on the left side using the same procedure. No cografting with the LGE was done as it could not be found in the 6 week-old fetus. The postoperative course was smooth. The patient was discharged 1 week later, and cyclosporin-A was maintained for 3 months after the second transplantation.

Results

Behavioral changes: The UPDRS scores of the patient before and after the neural transplantation were scored. During the "off" phase, the score was 128 before transplantation. Three and six months after transplantation, scores improved to 68 and 62, respectively. At 22 months post transplantation, the UPDRS score improved to 24. The patient had little motor fluctuation, no freezing or postural instability, no tremor of limbs, and no difficulty initiating movements. Six months after the last transplantation, he returned to work on his farm as he did 4 years ago when he last time worked there. His L-dopa dosage was kept at 150 mg three times a day during daytime and 50 mg after dinner.

The score of Activity of Daily Living, a subcomponent of UPDRS, during "off" phase was 35 before transplantation and improved to 17 at 3 months, 18 at 6 months, and 10 at 22 months after transplantation. During "on" phase, the scores were 12 before transplantation and 10, 9, and 7 at 3, 6 and 22 months, respectively, after transplantation.

[18F] 6-fluoro-L-dopa PET: The values of ki of the [18F] 6-fluoro-L-dopa in the caudate and the putamen 1 month before and 21 months after transplantation were measured. The ki in the right caudate decreased from 0.0060 to 0.0051 (15% decrease), and from 0.0063 to 0.0060 (5% decrease) in the left caudate. The caudate did not contain any fetal VM grafts. In contrast, the ki of the putamen was increased on both sides; on the right side, which had VM and LGE cografts, the ki increased from 0.0021 to 0.0030 (43% improvement); on the left side, which had VM grafting only, the ki increased from 0.0027 to 0.0035 (30% improvement). A previous report demonstrated that striatal dopaminergic terminals continuously degenerated over time in PD patients [12]. Accordingly, dopaminergic terminals in both caudate and putamen of our patient also degenerated over the 21-month period after transplantation. The actual increment of ki in the putamen, after adding the decrement of ki in the caudate, was 58% in the right and 35% in the left. Thus, the cografted side had a larger ki increment (58% vs. 35%).

Discussion

Our results, inferred from the clinical and PET data, show that fetal VM transplanted in the putamen of PD patients can survive and produce a positive effect for 21 months. Cografting fetal VM with fetal LGE appears to have enhanced the survival of the grafted dopamine neurons in the putamen of this patient.

[18F] 6-fluoro-L-dopa can be actively uptaken by dopamine neurons [2, 3, 12]. The ki of [18F] 6-fluoro-L-dopa is reported to be proportional to the number of dopamine neurons and the density of DA neurites. For a PD patient in the early stages, a loss of [18F] 6-fluoro-L-dopa uptake is noted in the posterior putamen. As the disease progresses, the decrease in [18F] 6-fluoro-L-dopa uptake extends from the posterior putamen to the anterior putamen. In patients with stage 4 in the Hoehn and Yahr Scale, most of the [18F] 6-fluoro-L-dopa uptake in the entire putamen is markedly decreased [2, 3, 12]. The pre-operative [18F] 6-fluoro-L-dopa PET in our patient is in agreement with this. The ki of the putamen in our patient was about 31% of age-matched normal values. In contrast, the averaged ki of the caudate for our patient was 67% of normal.

It has been reported that [18F] 6-fluoro-L-dopa PET imaging is a useful tool with which to follow the survival of the transplanted dopamine neurons. The ki of [18F] 6-fluoro-L-dopa was proportional to the number of the surviving dopamine neuron post transplantation, which was evaluated by subsequent histology [3]. For our patient, the ki for the [18F] 6-fluoro-L-dopa increased in the putamen bilaterally, suggesting survival of the grafted dopamine neurons 21 months after the transplantation.

The ki for [18F] 6-fluoro-L-dopa examined 1 month before and 21 months after transplantation showed a greater increment on the cografted side (the right, 58%) than the non-cografted side (the left, 35%). To our knowledge, this is the first human cografting case ever with follow-up [18F] 6-fluoro-L-dopa PET data, reported in the literature.

Because of the low survival rate of grafted dopamine neurons, more than 2 fetal VMs have been used previously for each side of the putamen to obtain symptomatic improvement [3]. In our patient, only one embryonic VM was transplanted into each putamen. The patient's symptoms of akinesia, rigidity, tremor and postural instability were all improved. The UPDRS score decreased from 128 to 24. This symptomatic improvement is not related to an increase in L-dopa therapy since the total dosage actually decreased, from 900 mg to 500 mg per day. Therefore, the symptomatic improvement was probably due to the grafted fetal dopamine tissue, especially in view of the substantial increase in ki (58% increase) on the cografted side. The ki for the [18F] 6-fluoro-L-dopa before the transplantation appears not to be markedly influenced by pallidotomy, since the ki in the putamen was similar on the pallidotomied and non-pallidotomied sides (0.0021 vs. 0.0027).

In this case, the ki of the left putamen (non-cografted side) received only one fetal VM increased by 35% after 21-month transplantation. We did screening of infectious diseases 1 week before the elective abortion. The embryo obtained, we removed the brain, dissected out the VM and preserved the brain tissue in cold 1% glucose Lactate Ringer solution. Then we transplanted the fetal VM into PD patient within 3 hours. We believed fresh tissue transplantation might have a good survival of dopamine neurons as indicated by the increased ki value.

Our pallidotomied PD patient was treated successfully with fetal VM grafting. One year before fetal VM transplantation, the patient had a pallidotomy in the left side, which led only to a transient (7 day) improvement in symptoms, suggesting the pallidotomy was not successful. These results imply that a failed pallidotomy does not completely destroy the outflow pathway from the internal globus pallidus to the thalamus and that PD patients who have an unsuccessful pallidotomy can subsequently be treated with neural transplantation.

In conclusion, we confirm and extend the concept that PD patients can be effectively treated by transplanting fetal VM. Cografting the fetal VM with the LGE may enhance the survival of the grafted dopamine tissue, and improve the efficacy of such grafts in PD.

Acknowledgment

The authors acknowledge Dr. A. Jon Stoessl, and Dr. Donald. B. Calne, Neurodegenerative Disorders Center, University of British Columbia, Vancouver, Canada, for their helping in fluro-dopa PET examination and data analysis. This study was supported by grants (NSC89-2314-B016-075; NSC89-2320-B016-129; NSC88-2314-B-016-012-M20; NSC87-2314-B016-067-M20; NSC86-2314-B-016-025-M20) from the National Science Council, Taiwan.

References

1. Freeman TB, Olanow CW, Hauser RA, Nauert GM, Smith DA, Borlongan CV, Sanberg PR, Holt DA, Kordower JH, Vingerhoets FJG, Snow BJ, Calne D, Gauger LL (1995) Bilateral fetal nigral transplantation into the postcommissural putamen in Parkinson's disease. Ann Neurol 38: 379–388
2. Hall H, Farde L, Halldin C, Hogberg T, Larsson S, Sedvall G (1992) Imaging of dopamine receptors using PET and SPECT. Neurochem Int 20: 329S–333S
3. Kordower JH, Freeman TB, Snow BJ, Vingerhoets FJG,

Mufson EJ, Sanberg PR, Hauser RA, Smith DA, Nauert GM, Perl DP, Olanow CW (1995) Neuropathological evidence of graft survival and striatal reinnervation after the transplantation of fetal mesencephalic tissue in a patient with Parkinson's disease. N Engl J Med 332: 1118–1124
4. Lindvall O, Brundin P (1990) Grafts of fetal dopamine neurons survive and improve motor function in Parkinson's disease. Science 247: 574–575
5. Madrazo I, Drucker-Colin R (1987) Open microsurgical autograft of adrenal medulla to the right caudate nucleus in two patients with intractable Parkinson's disease. N Engl J Med 316: 831–834
6. Plunkett RJ, Weber RJ (1988) Stereotaxic implantation of dispersed cell suspensions into brain: A systemic appraisal of cell placement and survival. J Neurosurg 69: 228–233
7. Science, Editorial (1992) Fetal transplants show promise. Science 257: 868
8. Sinclair SR, Svendsen CN, Torres EM, Martin D, Fawcett JW, Dunnett SB (1996) GDNF enhances dopaminergic cell survival and fiber outgrowth in embryonic nigral grafts. Neuroreport 7: 2547–2552
9. Spencer DD, Robbins RJ, Naftolin F, Marek KL, Vollmer T, Leranth C, Roth RH, Price LH, Gjedde A, Bunney BS, Kimberlee JS, Elsworth JD, Kier EL, Makuch R, Hoffer PB, Redmond DE Jr (1992) Unilateral transplantation of human fetal mesencephalic tissue into the caudate nucleus of patients with Parkinson's Disease. N Engl J Med 327: 1541–1548
10. Stromberg I, Johansson M (1994) Re-initiated growth from mature ventral mesencephalon: an in oculo transplant study of nigrostriatal cografts. Exp Brain Res 101: 72–85
11. Vingerhoets FJ, Snow BJ, Schulzer M, Morrison S, Ruth TJ, Holden JE, Cooper S, Calne DB (1994) Reproducibility of fluorine-18-6-fluorodopa positron emission tomography in normal human subjects. J Nucl Med 35: 18–24
12. Vingerhoets FJG, Snow BJ, Lee CS, Schulzer M, Mak E, Calne DB (1994) Longitudinal fluorodopa positron emission tomographyic studies of the evolution of idiopathic parkinsonism. Ann Neurol 36: 759–764
13. Wang Y, Wang SD, Lin SZ, Liu JC (1994) Restoration of dopamine overflow and clearance from the 6-hydroxydopamine lesioned rat striatum reinnervated by fetal mesencephalic grafts. J Pharm Exp Thera 270: 814–821
14. Widner H, Tetrud J, Rehncrona S (1992) Bilateral fetal mesencephalic grafting in two patients with parkinsonian induced by 1-methyl-4-phenyl-1,2,3,6-tetrahydropyridine. N Engl J Med 327: 1556–1563
15. Yurek DM, Collier TJ, Sladek JR (1990) Embryonic mesencephalic and striatal co-grafts: development of grafted dopamine neurons and functional recovery. Exp Neurol 109: 191–199

Correspondence: Shinn-Zong Lin, M.D., Ph.D., Neuro-Medical Scientific Center, Tzu-Chi General Hospital, Tzu-Chi University 707, Sec. 3, Chung Yang RD. Hualien, Taiwan, 970. E-mail: szlin@tzuchi.org.tw

Acta Neurochir (2003) [Suppl] 87: 163–168

Effect of subthalamic lesion with kainic acid on the neuronal activities of the basal ganglia of rat Parkinsonian models with 6-hydroxydopamine

J. W. Chang[1], **J. S. Yang**[2], **M. F. Jeon**[1], **B. H. Lee**[1], and **S. S. Chung**[1]

[1] Department of Neurosurgery, Medical Research Center, Brain Korea 21 project for Medical Science & Brain Research Institute, Yonsei University College of Medicine, Seoul, Korea
[2] Graduate School of East-West Science, Kyung Hee University, Seoul, Korea

Summary

The aim of the present study was to investigate the alteration of neuronal activities in the substantia nigra pars reticulata (SNpr) and globus pallidus (GP), after ipsilateral STN lesioning by kainic acid in the rat hemi-parkinsonian 6-hydroxydopamine (6-OHDA) model. In various rat parkinson's disease (PD) models, an increase in the SNpr firing rate was observed, despite the occurrence of bursting patterns, and subthalamic lesion was found to reduce the mean firing rates and the percentage of bursting neurons in the SNpr. However, the relative proportion of bursting neurons, among all GP neurons, was slightly increased as a result of the subthalamic lesion. The significance of bursting activity in the SNpr and GP remains obscure. Further study is necessary to elucidate the pathophysiological mechanism behind parkinson's disease.

Keywords: 6-hydroxydopamine; Parkinson's disease; subthalamus; basal ganglia.

Abbreviations

PD Parkinson's disease; *6-OHDA* 6-hydroxydopamine; *STN* subthalamic nucleus; *MFB* medial forebrain bundle; *SN* substantia nigra; *L-DOPA* L-3,4-dihydroxyphenylalanine; *DA* dopamine.

Introduction

Current concepts regarding the functional organization of the basal ganglia, based largely on experimental findings in primate models, point to subthalamic nucleus (STN) hyperactivity as being a major factor in the pathophysiology of Parkinson's disease (PD) [12]. It is hypothesized that the hyperactivity of STN neurons in PD results from the loss of dopamine in the striatum, leading to a reduction in the activity of the inhibitory pallido-subthalamic pathway. The STN has been shown to exert an excitatory influence on the output structures of the basal ganglia. In accordance with this hypothesis, high frequency stimulation of the STN, in patients suffering from a severe form of PD, has been observed to successfully alleviate the disabling akinesia, rigidity and tremor [4, 18, 19]. Nonetheless, in spite of numerous investigations, the mechanism behind PD remains obscure. However, recent reports have provided controversial results, such as increased or unchanged activity of the inhibitory pallido-subthalamic pathway [8, 14, 22, 24, 25]. Furthermore, a detailed analysis of spike trains, recorded from the basal ganglia nuclei of parkinsonian animals, reveals the importance of changes in the firing pattern, as well as changes in the firing rate. Moreover, most reports point to the existence of changes in the firing pattern, including increased variability of the interspike intervals (ISIs), a tendency of units to fire in bursts and to show periodic oscillations in firing rate [8, 27, 31]. However, the relative importance of these parameters remains obscure, partly because the changes seem to affect only a fraction of the sampled units, and because it is unclear whether changes in the firing pattern and firing rate occur in the same or in different units. The aim of the present study was to investigate the alteration of these neuronal activities in the basal ganglia (GP, globus pallidus; SNpr, substantia nigra pars reticula) after ipsilateral STN lesioning by kainic acid in the rat hemi-parkinsonian 6-hydroxydopamine (6-OHDA) model.

Material and methods

Surgical procedures for the medial forebrain bundle (MFB) and STN lesion

A total of 40 male Sprague-Dawley rats, weighing 200–250 g were used in the experiments. Rats were devided into 5 groups; (i) a control group (n = 10); (ii) a 6-OHDA lesion group (n = 10) having a lesion of dopaminergic neurons caused by the 6-OHDA; (iii) an STN lesion group (n = 7) including rats with kainic acid induced STN lesions; (iv) a 6-OHDA + STN lesion group (n = 7) comprising rats with a 6-OHDA lesion of dopaminergic neurons plus an STN lesion; (v) a 6-OHDA + STN sham group (n = 7) having received saline instead of kainic acid in the STN. All rats were housed in a temperature- and humidity-controlled room with a 12 hour light-dark cycle. The rats were anesthetized with a mixture of ketamine (75 mg/kg), acepromazine (0.75 mg/kg) and rompun (4 mg/kg). As previously described [10], 8 μg (free base weight) of 6-OHDA (RBI) in 0.2% ascorbic acid with 0.9% normal saline was infused in 2 μl at a rate of 0.5 μl/min at the following coordinates of MFB: AP −4.4 mm, ML 1.2 mm relative to bregma, and DV −7.5 mm from the dura. After injection of 6-OHDA, the cannula was left in place for 5 min before being slowly retracted. To prevent destruction of noradrenergic neurons, desipramine (12.5 mg/kg, i.p.) was administered 30 min prior to the infusion of 6-OHDA. The STN lesion was made with kainic acid (Sigma) 1 μg in 0.5 μl saline solution injected at a rate of 0.5 μl/min (total 1 μl) in the STN (coordinates: AP −3.7 mm, ML 2.5 mm relative to bregma, and DV −8.0 mm from the dura).

Microelectrode recording

For extracellular single unit recordings, rats were anesthetized with urethane (1.3 g/kg intraperitoneally). Glass micropipettes (impedence 7–10 Mohm at 100 Hz) were used for single unit recording. Microelectrodes were stereotaxically guided through a skull burr hole to the target coordinates (GP, AP, −0.92 mm; L, 3 mm and V, 6.0–6.6 mm; SNpr, AP, −5.3 mm; L, 2.4 mm and V, 7.4–8.0 mm; STN, AP, −3.7 mm; L, 2.5 mm and V, 7.4–7.8 mm). Electrical signals were passed through a DAM80 preamplifier (WPI, Sarasota, FL, USA) in bridge mode, and amplified signals were monitored with an oscilloscope and audio monitor. The stored signal was converted to square wave pulses with the aid of a window discriminator (WPI, Sarasota, FL, USA) and fed onto an A/D converter and personal computer. The ISIs allowed the evaluation of the unit's degree of burstiness, following an algorithm described by Hutchison *et al.* (1997, 1998). Bursting cells were defined as having a score of more than 10, based on the reciprocal of the modal interval divided by the mean firing rate. The autocorrelogram evaluated the rhythmicity and the cut-off points of the rhythmicity as being 0.95. To characterize and study the distribution of the firing patterns, the discharge pattern was classified according to the number of clearly identifiable peaks in the autocorrelogram, as described in a previous report [8].

Analysis of lesions and data

At the end of the extracellular single unit recording, neurons were identified by their stereotactic location and by the histological location of the electrode tip, following iontophoresis of Pontamine sky blue from the recording electrode (−18 μA for 20 min). Rats were anesthetized and transcardially perfused with 125 ml normal saline, followed by 250 ml ice-cold 4% paraformaldehyde. Brains were removed, postfixed for 10 hours, and placed in a 30% sucrose solution until equilibrated. 20 μm sections were cut, frozen and immunoreacted with a primary, polyclonal antibody against rat TH (Pelfreeze, Rogers, AK) at a dilution of 1:750 and a biotinylated goat anti-rabbit IgG (Vector Labs, Burlingame, CA) secondary antibody. The STN lesions and the localization of recorded basal ganglia nuclei were studied in 20 μm sections stained with cresyl violet.

Data were reported as mean ± SEM. Statistical significance levels were calculated using the ANOVA method. The statistical significance level was set at $p < 0.05$. All statistical analyses were performed using SPSS version 9.0 (SPSS Inc., Chicago, IL, USA).

Results

Histological findings of rat PD models

The extent and location of the lesions caused by the 6-OHDA were confirmed by assessing the loss of TH-immunoreactive cells and fibers in the substantia nigra pars compacta (SNpc) and striatum in the rat parkinsonian model with 6-OHDA (Fig. 1A). In additon, the ventral tegmental area (VTA) was also lesioned on the same side in most of the rats. The STN lesions were also evaluated following the completion of the experiments, and revealed evidence of local gliosis at the level of the STN (Fig. 1B). The localization of the recorded basal ganglia nuclei (GP, SNpr) were confirmed by staining with cresyl violet (Fig. 1C).

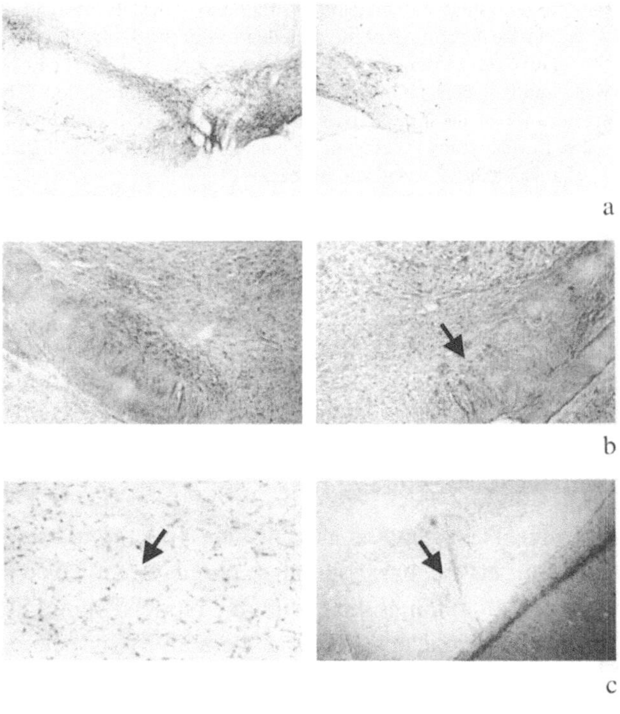

Fig. 1. (a) Immunohistochemistry of tyrosine hydroxylase (*TH*) showing the total degeneration of dopamine fiers in the striatum and dopamine cell bodies in the SNpc on the 6-OHDA injected side (right) compared to normal side (left). (b) Cresyl violet-stained sections illustrating a unilateral kainic acid lesion in the subthalamic nucleus. Arrow indicates the location of the lesion. (c) Photomicrograph showing the Pontamine Sky Blue mark corresponding to a neuron recorded at the end of a track in the globus pallidus (left) and (right) pars reticula of substantia nigra. Magnification, ×40

Table 1. *The general results from basal ganglia (GP, SNpr, STN) neurons recorded in normal rats, and in rats with SNpc, STN and SNc + STN lesions*

	GP			SNpr			STN		
	Neurons (*n*)	Mean firing rate (Hz)	Interspike interval (sec)	Neurons (*n*)	Mean firing rate (Hz)	Interspike interval (sec)	Neurons (*n*)	Mean firing rate (Hz)	Interspike interval (sec)
Control	34	19 ± 1.6	0.20 ± 0.012	42	20 ± 1.3	0.04 ± 0.019	27	11 ± 1.0	0.15 ± 0.053
PD	42	16 ± 1.3*	0.04 ± 0.004	62	27 ± 1.6*	0.06 ± 0.024	49	20 ± 1.8*	0.44 ± 0.106
STN	35	16 ± 1.2	0.05 ± 0.005	33	19 ± 1.6	0.05 ± 0.005			
PD + STN	44	18 ± 1.6	0.04 ± 0.006	36	24 ± 2.2	0.02 ± 0.007			
PD + STNsham	30	16 ± 1.7	0.04 ± 0.005	35	26 ± 1.7	0.04 ± 0.020	37	20 ± 2.0	0.36 ± 0.090

The values means ± SEM. * P < 0.05 in comparison with values from normal animals.

Table 2. *The proportions of basal ganglia (GP, SNpr, STN) neurons, according to their firing pattern, recorded in normal rats, and in rats with PD, STN and PD + STN lesions*

		Control	PD	STN	PD + STN	PD + STNsham
GP	Regular (%)	82	83	77	66	83
	Burst (%)	18	17	23	34	16
SNpr	Regular (%)	71	56	67	80	66
	Burst (%)	29	44	33	20	32
STN	Regular (%)	70	62			64
	Burst (%)	30	38			36

Effects of STN lesion on firing rate and firing patterns of basal ganglia nuclei in rat PD models

In normal unlesioned rats, the mean firing rates of neurons were as follows; 19 ± 1.6 Hz in the GP (n = 34), 20 ± 1.3 Hz in the SNpr (n = 42), 11 ± 1.0 Hz in the STN (n = 27). Compared to the normal control rats, rat PD models with 6-OHDA exhibited a significant increase in mean firing rates in the STN (20 ± 1.8 Hz) and SNpr (27 ± 1.6 Hz). However, the mean firing rate in the GP (16 ± 1.3 Hz) decreased ($p < 0.05$). In the STN lesioned rat PD models, the mean firing rates in the SNpr (24 ± 2.2 Hz) were reduced and the mean firing rates in the GP (18 ± 1.6 Hz) were increased ($p < 0.05$). There was no statistically significant difference in the mean firing rates in the sham STN lesioned PD rats (26 ± 1.7 Hz) ($p > 0.05$). Analysis of ISI histogram, raster displays and autocorrelograms was used to classify the discharge patterns of the basal ganglia nuclei. We classified the discharge patterns of the SNpr and GP as involving bursting or non bursting neurons according to an algorithm described by Hutchison *et al.* (1997, 1998). The non-bursting neurons are composed of tonic neurons and mixed tonic-bursting neurons. Bursting neurons represented 30% of the total neurons in the STN in the control rats and 38% in PD rats. Of the SNpr neurons, bursting neurons represented 29% of the total in the control rats and 44% in PD rats. Following STN lesioning for the rat PD models, the percentage of bursting neurons decreased by 10% ($p < 0.05$). However, STN lesioned control rats and the sham STN lesioned rat PD models did not show a significant change in the percentage of the bursting neurons in SNpr ($p > 0.05$). In the group of rats with 6-OHDA lesions, the bursting neurons in the GP represented 17%, which was similar to that of normal rats. In the group of rats containing two lesions, SNpc lesions and STN lesions, the percentage (34%) of the bursting neurons in the GP was increased significantly compared to that of the normal group (18%). The percentage (23%) of the bursting neurons in the GP in the group of rats containing only STN lesions, was also increased.

Discussion

A current model of the motor pathway postulates that D1 dopamine receptors on striatonigral neurons act to increase the activity of the inhibitory GABAergic striatonigral projection [12, 14]. Thus, as a result of the dopaminergic denervation of the striatum in rat PD models, striatal GABAergic neurons projecting to SNpr should become underactive, leading to less inhibition of this output nucleus. This in turn should lead to overactivity of the GABAergic SNpr neurons, overactivation which would be further amplified by increased excitatory input from the STN. Our study produced data which were in accordance with these predictions. We confirmed that rat PD models had increased firing rates in the SNpr (27 ± 1.6 Hz), STN (20 ± 1.8 Hz) and that STN lesioning with kainic acid in rat PD models reduced the hyperactivity of SNpr neurons, in rat parkinsonian models, to a normal level (24 ± 2.2 Hz). In addition, there was no statistically significant difference in the mean firing rates in the sham STN lesioned PD rats (26 ± 1.7 Hz) ($p > 0.05$). According to the traditional pathophysiological model of parkinsonism, depletion of dopaminergic neurons in the SNc induces a decrease in the firing rate of GP neurons [2]. Albin [1] reported that increases in STN neuronal activity are likely due to a decrease in the level of GABA neurotransmission between GP neurons and their target STN neurons [1]. Our present results confirmed this decrease in the firing rate of GP neurons. In the 6-OHDA lesioned group having STN lesions (18 ± 1.6 Hz), the mean firing rate of GP neurons was nomalized in comparison to that of rats with only 6-OHDA lesions (16 ± 1.3 Hz). We observed that rat PD models had more bursting neurons (44%) in the SNpr than the control rats (29%). Following STN lesioning for the rat PD models, the percentage of bursting neurons decreased by 10% ($p < 0.05$). However, the STN lesioned control rats and the sham STN lesioned PD rats did not show a significant change in the percentage of bursting neurons in the SNpr ($p > 0.05$). However, many reports showed controversial results. Kreiss *et al.* [17] reported that rat PD models had more regular neurons in the STN than the control rats. Rohfts *et al.* [27] reported that SNpr neuronal activity in rat PD models was significantly lower than the respective activity in the control groups and, in contrast, contralateral SNpr neuronal activity in rat PD models was significantly higher than the respective activity in the contralateral SNpr neuronal activity of control rats. In addition to the significant reduction in the mean firing rate, they also reported that many neurons in the SNpr on the lesioned side exhibited a bursting pattern. MacLeod *et al.* [20] suggested that the continuous bursting type of firing could be an indication of a compensatory physiological mechanism for the loss of spontaneous firing, and that the decreased mean firing rate of ipsilateral SNpr neurons, observed in rats within 2 weeks post-lesion, can be explained by assuming a postsynaptic inhibitory action of DA in the striatum. However, our data did not seem to support these suggestions. All of our rat PD models with 6-OHDA clearly demonstrated an increase in the mean firing rates in the SNpr. In addition, the results of our rat PD models, having STN lesions, further supports the theory that the mean firing rates in the SNpr are reduced by the presence of STN lesions.

As regards the bursting pattern, a similar increase to that we observed in the proportion of bursty SNpr neurons in the ipsilateral SNpr of 6-OHDA lesioned rat PD models has previously been reported [20, 30]. We observed a tonic discharge pattern of both regular and irregular type non-bursting neurons in the SNpr of rat PD models, as has been previously reported [8, 31]. In the rat PD models, an increase in the SNpr firing rate was observed, despite the occurrence of bursting patterns and subthalamic lesions, which reduced the mean firing rates and the percentage of bursting neurons in the SNpr. However, there have been reports that the basal firing rates of STN neurons in rat PD models do not substantially differ from those observed in control rats [15], although pallidal lesions did significantly increase these firing rates [29]. In both cases, the incidence of bursting patterns was enhanced when compared to control rats. Our data reinforce the view that the STN provides increased excitatory input to basal ganglia output structures, as has been demonstrated in both rats and monkeys [13, 26]. Numerous electrophysiological studies have shown that variations in the mean discharge rate of GP neurons are associated with a change in firing pattern [14, 24]. Ni *et al.* [23] reported that STN lesions led to the disappearance of burst firing cells when compared to normal rats, and to a normalization of the firing pattern in rats with previous SNc lesions. However, the results of the present study showed no significant change in firing pattern in the rats with SNc lesions. In the group having both STN and SNc lesions, STN lesions were seen to lead to a slight increase in burst firing cells in comparison to 6-OHDA lesion rats.

The STN receives its main cortical afferents from the sensory motor cortex and enhances the selectivity of cortical information impinging on basal ganglia output structures [28]. This role would tend to be more marked in unanesthetized animals, since cortical areas may generate tonic, and especially phasic activity, in STN neurons, arising from normal motor activity. We performed this experiment using urethane-anesthetized rats. Since general anesthesia with urethane could have an influence on the firing rates of the animal, local anesthesia with muscle block had previously been recommended [17, 29]. However, Hassani et al. [14] argued that urethane did not prevent the stimulation of STN neurons in the rat PD model with 6-OHDA by drugs as assessed by the electrophysiological response. Furthermore, many reports have been published in which experiments were performed using general anesthesia with urethane or chloral hydrate, as in this experiment [21, 23, 31].

Acknowledgments

This work was supported by grant No. (2000-2-20600-001-3) from the Basic Research Program of the Korea Science & Engineering Foundation.

References

1. Albin RL, Young AB, Penney JB (1989) The functional anatomy of basal ganglia disorders. Trends Neuroci 12: 366–375
2. Alexander GE, Crutcher MD, DeLong MR (1990) Basal ganglia thalamocortical circuits: parallel substrates or motor, oculomotor, 'prefrontal' and 'limbic' functions. Prog Brain Res 85: 119–146
3. Aziz TZ, Peggs D, Sambrook MA, Crossman AR (1991) Lesion of the subthalamic nucleus for the alleviation of 1-methyl-4-phenyl-1,2,3,6-tetrahydropyridine (MPTP)-induced parkinsonism in the primate. Mov Disord 6: 288–292
4. Benabid AL, Koudsie A, Benazzouz A, Vercueil L, Fraix V, Chabardes S, LeBas JF, Pollak P (2001) Deep brain stimulation of the corpus luysi (subthalamic nucleus) and other targets in Parkinson's disease. Extension to new indications such as dystonia and epilepsy. J Neurol 248 [Suppl] 3: III/37–III/47
5. Benazzouz A, Gao DM, Ni ZG, Piallat B, Bouali-Benazzouz R, Benabid AL (2000) Effect of high-frequency stimulation of the subthalamic nucleus on the neuronal activities of the substantia nigra pars reticulata and ventrolateral nucleus of the thalamus in the rat. Neuroscience 99: 289–295
6. Benazzouz A, Gross C, Geger J, Boraud T, Bioulac B (1993) Reversal of rigidity and improvement in motor performance by subthalamic high-frequency stimulation in MPTP-treated monkeys. Eur J Neurosci 5: 382–389
7. Bergman H, Wichmann T, DeLong MR (1990) Reversal of experimental parkinsonism by lesions of the subthalamic nucleus. Science 249: 1436–1438
8. Burbaud P, Gross C, Benazzouz A, Coussemacq M, Bioulac B (1995) Reduction of apomorphine-induced rotational behaviour by subthalamic lesion in 6-OHDA lesioned rats is associated with a normalization of firing rate and discharge pattern of pars reticulata neurons. Exp Brain Res 105: 48–58
9. Chang JW, Wachtel SR, Young D, Kang UJ (1999) Biochemical and anatomical characterization of forepaw adjusting steps in rat models of parkinsons disease: studies on medial forebrain bundle and striatal lesions. Neuroscience 88: 617–628
10. Chang JW, Lee BH, Lee MS, Chang JH, Park YG, Chung SS (2001) Microelectrode Recording-Guided Deep Brain Stimulation in Patients with Movement Disorders (The First Trial in Korea). In: Lultas-Ilinsky K, Ilinsky IA (eds) Basal ganglia and thalamus in health and movement disorders. Kluwer Academic/ Plenum Publishers, New York, pp 341–347
11. Delfs JM, Ciaramitaro VM, Parry TJ, Chesselet MF (1995) Subthalamic nucleus lesions: widespread effects on changes in gene expression induced by nigrostriatal dopamine depletion in rat. J Neurosci 15: 6562–6575
12. Delong MR (1990) Primate model of movement disorders of basal ganglia origin. Trends Neurosci 13: 281–285
13. Hamada I, DeLong MR (1992) Excitotoxic acid lesions of the primate subthalamic nucleus result in transient dyskinesias of the contralateral limbs. J Neurophysiol 68: 1850–1858
14. Hassani OK, Mouroux M, Feger J (1996) Increased subthalamic neuronal activity after nigral dopaminergic lesion independent of disinhibition via the globus pallidus. Neuroscience 72: 105–115
15. Hollerman JR, Grace AA (1992) Subthalamic nucleus firing in the 6-OHDA treated rat: basal activity and responses to haloperidol. Brain Res 590: 291–299
16. Hutchinson WD, Allan RJ, Opitz H, Levy R, Dostrovsky JO, Lang AE, Lozano AM (1998) Neurophysiological identification of the subthalamic nucleus in surgery for Parkinson's disease. Annals Neurol 44: 622–628
17. Kreiss DS, Mastropietro CW, Rawji SS, Walters JR (1997) The response of subthalamic nucleus neurons to dopamine receptor stimulation in a rodent model of Parkinson's disease. J Neurosci 17: 6807–6819
18. Kumar R, Lozano AM, Kim YJ, Hutchison WD, Sime E, Halket E, Lang AE (1998) Double-blind evaluation of subthalamic nucleus deep brain stimulation in advanced Parkinson's disease. Neurology 51: 850–855
19. Limousin P, Krack P, Pollak P, Benazzouz A, Ardouin C, Hoffmann D, Benabid AL (1998) Electrical stimulation of the subthalamic nucleus in advanced Parkinson's disease. New Eng J Med 339: 1105–1111
20. MacLeod NK, Ryman A, Arbuthnott GW (1990) Electrophysiological properties of nigrothalamic neurons after 6-hydroxytryptamine lesions in the rat. Neuroscience 38: 447–456
21. Meissner W, Reum T, Paul G, Harnack D, Sohr R, Morgenstern R, Kupsch A (2001) Striatal dopaminergic metabolism is increased by deep brain stimulation of the subthalamic nucleus in 6-hydroxydopamine lesioned rats. Neurosci Lett 303: 165–168
22. Nakao N, Ogura M, Nakai K, Itakura T (1998) Intrastriatal mesencephalic grafts affect neuronal activity in basal ganglia nuclei and their target structures in a rat model of Parkinson's disease. J Neurosci 18: 1806–1817
23. Ni Z, Bouali-Benazzouz R, Gao Dongming, Benabid AL, Benazzouz A (2000) Changes in the firing pattern of globus pallidus neurons after the generation of nigrostriatal pathway are mediated by the subthalamic nucleus in the rat. Eur J Neurosci 12: 4338–4344
24. Pan HS, Walters JR (1988) Unilateral lesion of the nigrostriatal pathway decrease the firing rate and alters the firing pattern of globus pallidus neurons in the rat. Synapse 2: 650–656
25. Porter RHP, Greene JG, Higgins Jr DS, Greenamyre JT (1994) Polysynaptic regulation of glutamate receptors and mitochondrial enzyme activities in the basal ganglia of rats with unilateral dopamine depletion. J Neurosi 14: 7192–7199
26. Robledo P, Feger J (1990) Excitatory influence of rat subthalamic nucleus to substantia nigra pars reticulata and the pallidal complex: electrophysiological data. Brain Res 518: 47–54

27. Rohlfs A, Nikkhah G, Rosenthal C, Rundfeldt C, Brandis A, Samii M, Löscher W (1997) Hemispheric asymmetries in spontaneous firing characteristics of substantia nigra pars reticulata neurons following unilateral 6-hydroxydopamine lesion of the rat nigrostriatal pathway. Brain Res 761: 352–356
28. Ryan LJ, Clark KB (1992) Alteration of neuronal responses in the subthalamic nucleus following globus pallidus and neostratal lesions in rats. Brain Res Bull 29: 319–327
29. Ruskin DN, Bergstrom DA, Mastropietro CW, Twery MJ, Walters JR (1999) Dopamine agonist-mediated rotation in rats with unilateral nigrostriatal lesions is not dependent on net inhibitions in basal ganglia output nuclei. Neuroscience 91: 935–946
30. Sanderson P, Mavoungou R, Albe-Fessard D (1986) Changes in substantia nigra pars reticulata activity following lesions of the substantia nigra pars compacta. Neurosci Lett 67: 25–30
31. Tseng KY, Kasanetz F, Kargieman L, Pazo JH, Murer MG, Riquelme LA (2001) Subthalamic nucleus lesions reduce low frequency oscillatory firing of substantia nigra pars reticulata neurons in a rat model of Parkinson's disease. Brain Res 904: 93–103

Correspondence: Jin Woo Chang, M.D., Department of Neurosurgery, Yonsei University College of Medicine, CPO Box 8044, Seoul, Korea. E-mail: jchang@yumc.yonsei.ac.kr

Acta Neurochir (2003) [Suppl] 87: 169–174

Neural stem/progenitor cells survive and differentiate better in PD rats than in normal rats

Z. H. Sun[1], Y. L. Lai[1], W. W. Zeng[1], D. Zhao[1], H. C. Zuo[2], and Z. P. Xie[1]

[1] Institute of Neuroscience of Department of Biological Sciences and Biotechnology, Tsinghua University, Beijing, China
[2] Neurosurgical Center of Medical School, Tsinghua University, Beijing, China

Summary

To investigate the effects of grafted neural stem/mesencephalic progenitor cells (NSCs/MP) on rotational behavior of Parkinson's disease (PD) rats and the influence of intracerebral environment on NSCs/MP, we observed the survival and differentiation of NSCs/MP transplanted into 6-hydroxydopamine (6-OHDA)-lesioned and intact striatums. NSCs/MP were prepared from E_{11-15} rats and proliferated in serum-free medium with bFGF for several weeks. One day after being primed with serum/dbcAMP to differentiate, cell suspensions were grafted into 6-OHDA-lesioned and intact striatums respectively. It had been found that NSCs/MP were able to survive better and differentiate into more tyrosine hydroxylase (TH)-positive neurons in 6-OHDA-lesioned striatums than in intact ones, and apomorphine-induced rotations were obviously attenuated in MP graft models. The data suggested that NSCs/MP tend to survive and differentiate into TH-positive neurons in 6-OHDA-lesioned striatums. The data demonstrated that striatums in which DAergic terminals are destroyed by 6-OHDA undergo some changes and thus provide more appropriate conditions for NSCs/MP to differentiate into mature DAergic neurons. Furthermore, the finding that MP had greater relieving effects on rotational behavior than NSCs suggests that NSCs could not be used in clinical therapy of PD unless being induced into MP in vitro before transplantation.

Keywords: Neural stem cells; mesencephalic progenitors; 6-OHDA-lesioned striatum; TH-positive neurons.

Introduction

The etiology of many neurodegenerative diseases has been identified in recent years. In the past years, primary cell therapy has been brought forward as possible strategy for neurodegenerative diseases. However, insufficient amounts of donor cells limited the usage. Recently, basic fibroblast growth factor (bFGF)-responsive neural stem/mesencephalic progenitors (NSCs/MP) have been obtained from mammalian adult and embryonic brains [1, 2, 6–8, 10, 11]. They are self-renewable and multipotential cells and can generate all major cell types of the adult central nervous system, including neurons and glia. It has been approved that NSCs/MP grafted into brains can differentiate into various phenotype neurons, matching their target areas but apart from their primary differentiation fate. NSCs/MP can survive, migrate and integrate into host targets after being transplanted into adult brain [2, 11]. By means of their biological properties, NSCs/MP would be good candidates for multiple cell-based therapies for neural disease. Previous studies provided the evidence that striatum bearing unilateral 6-hydroxydopamine (6-OHDA) lesion was more effective on survival and differentiation of primary dopaminergic (DAergic) neurons and PC12 cells in vitro [4]. Some neurotrophic growth factors may be involved in this process [4, 5, 12]. In the present study, NSCs/MP from embryonic cortex and mesencephalon propagated with bFGF were grafted into intact and 6-OHDA-lesioned striatums, among which grafted cells labeled with fluorescent dye and tyrosine hydroxylase (TH)-immunopositive neurons were examined after transplantation. Furthermore, rotation behaviors were tested continually after transplantation. We tried to investigate the effects of NSCs/MP on Parkinson's disease (PD) and whether striatum destroyed by 6-OHDA might provide a more suitable environment for differentiation and maturation of grafted NSCs/MP in vivo.

Materials and methods

NSCs/MP' proliferation and expansion in vitro and labeling with DiI

Cortices and mesencephalon from E_{11-15} rat embryos were separated, mechanically dissociated, and plated at a final density of 5×10^5 cells/flask in 10 ml serum-free medium (consisting of a 1:1 mixture of F12 and DMEM containing 20 ng/ml bFGF) and incubated at 37°C in a 7.5% CO_2 condition. N_2/B_{27} (Gibco) were added to the medium while culturing MP. After the rat NSCs had been cultured for 5–7 days, the number of neurospheres increased rapidly. MP for transplantation were derived from primary neurospheres being cultured for 10 days in vitro. After the 5th generation of NSCs, neurospheres were then trypsinized off the bottom of the flask, pelleted, resuspended and dissociated mechanically until a homogenous single cell suspension was gained up to a concentration of 2×10^5/μl for neural transplantation.

Dissociate labeled NSCs/MP with 1,1′-Dioctadecy 1-3,3,3′,3′-tetramethylindocarbocyanine perchlorate (DiI, Biotium) [3]. DiI belongs to a family of cyanine dyes that are used to locate the transplanted cells. DiI absorbs maximally at 546 nm and has its maximum emission at 563 nm. Cells labeled with DiI were excited and viewed with filters appropriate for rhodamine fluorescent bright orange-red.

Preparation of 6-OHDA-lesioned rats and rotation behavior test [9, 12]

Female Sprague-Dawley rats (180–200 g) (n = 60) were housed with food and water ad lib. They received two stereotactic injections of 6-OHDA into the right striatum under anaesthesia at the following coordinates: (2.5 μl of 8 μg/μl 6-OHDA HBr in 0.9% saline with 0.2% ascorbate, all from Sigma) AP: −0.2, R: 3.5, V: 6 and AP: +1.8, R: 4.0, V: 6.5 according to Paxinos and Watson atlas. The injection rate was 1 μl/min and the microsyringe was rested for additional 5 mins before being retracted slowly. Four to six weeks after surgery, animals were given apomorphine (APO, Sigma, 0.1 mg/kg, s.c.) challenge and their rotation behavior was tested using an automated monitor system. The scores were collected over a 30 min period. Only animals exhibiting a mean ipsilateral rotation score of 7 or more full body turns/min were used in further study. The rotation behavior test was carried out at a period of 30 mins in 2, 4, 6 and 8 weeks after transplantation. Among them, normal and PD animals transplanted with DiI-labeled NSCs/MP were killed 4 weeks later. However, the others remained to be tested until 8 weeks when the morphologic examination was carried out.

Immunocytochemistry or Immunofluorescent staining of cultured cells and vibrosections

Cultured cells were fixed in 4% fresh paraformaldehyde in 0.1 M phosphate buffer saline. Perfused brains were immersed in the same solution overnight and coronal sections of 50 μm thick were prepared on a vibroslice microtome and collected serially in 0.1 M phosphate buffer. One in every four sections was processed for examining DiI-labeled and TH-immunofluorescent cells, respectively. The following mouse anti-rat monoclonal antibodies were used: TH, 1:10000 and MAP_2, 1:200 (all from Sigma) and nestin, 1:400 (Chemicon). The PV6002 (PowerVision) kit and 3,3′-diaminobenzidine (DAB) were used for visualizing the cultured cells. Briefly, cells were rinsed twice with PBS. Following that, cells were incubated for 30 mins at 37 °C with the anti-TH/MAP_2 antibody. Then cells were washed for 5 to 10 mins, followed by 30 mins' incubation at room temperature with goat anti-mouse IgG coupled to peroxidase. After secondary antibody polymer binding, peroxidase activity was developed using DAB chromagen and examined. The immunofluorescent staining was used to view the slices and nestin-labeled NSCs. Once the primary antibody was bound, the FITC-labeled secondary antibody was added and incubated at 37°C in dark for 1 hr. Fluorescence was observed subsequently.

Results

NSCs/MP propagation in vitro

After being cultured in serum-free medium supplemented with bFGF, both NSCs and MP propagated well and more than 85% NSCs/MP were nestin-immunopositive (Fig. 1A and 1B). The NSCs can be passaged through at least 5 generations and be maintained for more than 2 months, thus they can proliferate 10^6 times or more in vitro, which is enough for neural transplantation. Once the neurospheres/clones were dissociated mechanically and stuck to the P-D-lysine coated plate, the cells began to differentiate. About 20% differentiated cells revealed to be MAP-2 immunopositive neurons (Fig. 1C and 1D). Few NSCs differentiated into TH-positive neurons, but MP were able to differentiate into more TH-immunopositive neurons than NSCs (Fig. 1E and 1F).

Survival and differentiation of NSCs/MP after being transplanted into rat striatum

DiI-labeled NSCs/MP were implanted into the striatums of 3/2 rats in the control groups. The majority of NSCs/MP died 4 weeks later. About 1–2% transplanted NSCs/MP survived and existed only at the margin of the transplantation sites (Fig. 2A). The differentiated neurons accounted for only 1% of the surviving cells. The majority of NSCs survived in a good state after being implanted into PD rats and their differentiation is more active 4 weeks later. They spread not only in the center but also at the surrounding of the transplantation sites. The differentiated cells were about 5% (Fig. 2B). We could also examine few TH-immunopositive neurons 8 weeks later in the normal rats receiving transplantatation of unlabeled NSCs/MP. The number of TH-immunopositive neurons were 20/50 at the upmost. There was a significant difference between normal and PD rats ($P < 0.05$). Similar results were found in MP transplantation (Fig. 2C and 2D). At least 28 TH-immunopositive neurons were observed in the PD rats that had been transplanted with unlabeled NSCs 8 weeks later. Among them, 252 TH-immunopositive neurons were found in a PD model receiving NSCs grafts (Fig. 2E).

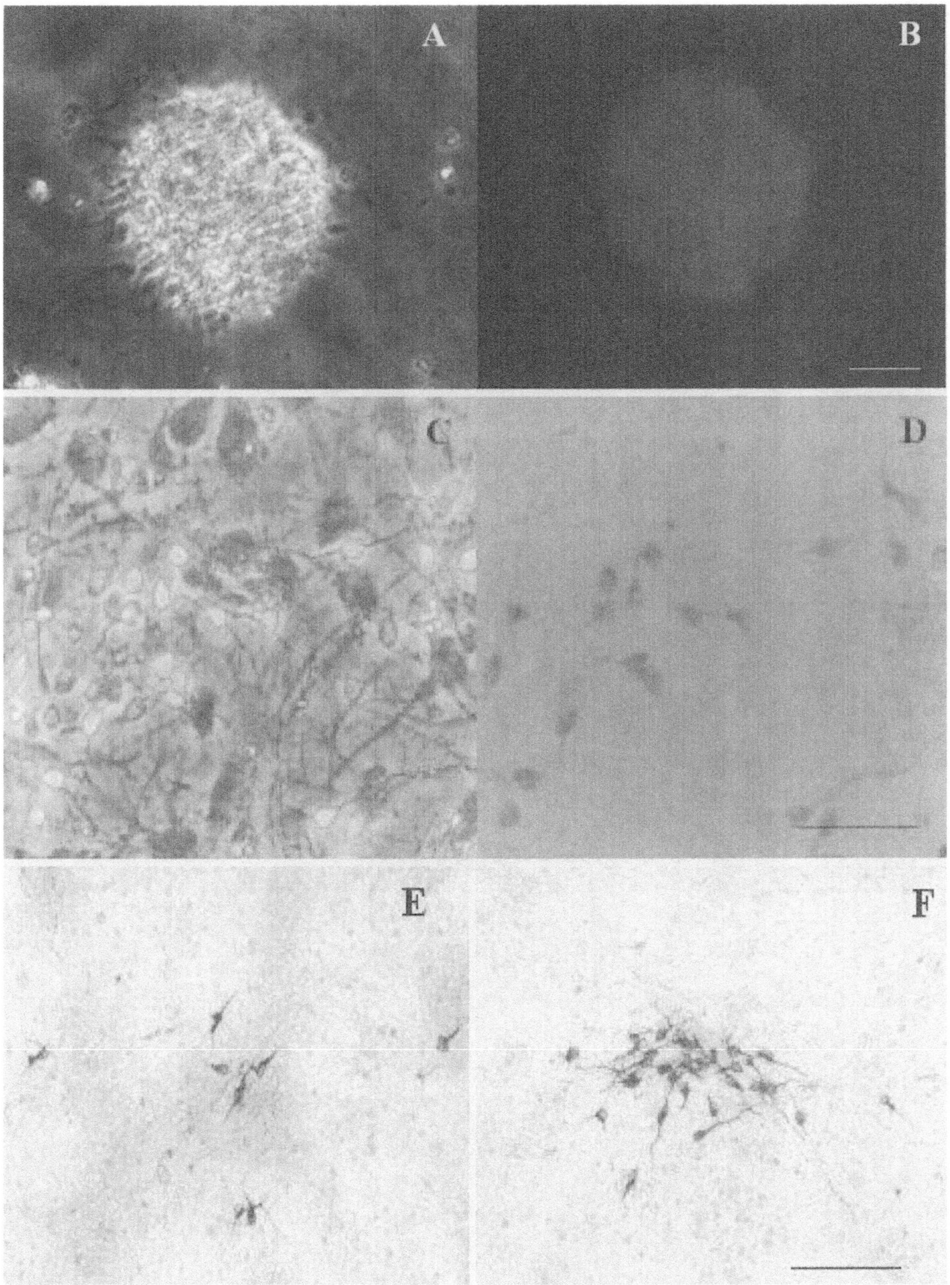

Fig. 1. Cortical cells grew into clones (A and B) after being cultured in serum-free medium in vitro. They could differentiate into MAP-positive neurons (C and D) after five days' culture on P-D-lysine coated plate supplemented with serum/dbcAMP. The neurospheres formed in vitro were photographed as shown in the phase-contrast image (A). They were nestin-positive as in B. Otherwise, the majority of neurospheres were primed with serum to differentiate (C), some of which differentiated into neurons (D, MAP-2-positive). TH-positive neurons also emerged (E and F). Few TH-positive neurons were included in the NSCs differentiation (E), however, MP differentiated into more TH-positive neurons in F (Scale bar: 100 μm)

The effects of NSCs/MP implanted into the striatum on the rotation behavior of PD rats

Most animals exhibited asymmetric turning behavior in response to APO (0.1 mg/kg, s.c.) challenge 4 weeks after they had received unilateral 6-OHDA lesions of the nigrostriatal pathway and the behavior was stable 6 weeks later. We equilibrated the two scores finished in two weeks and the average was considered the pre-graft control. NSCs/MP were trans-

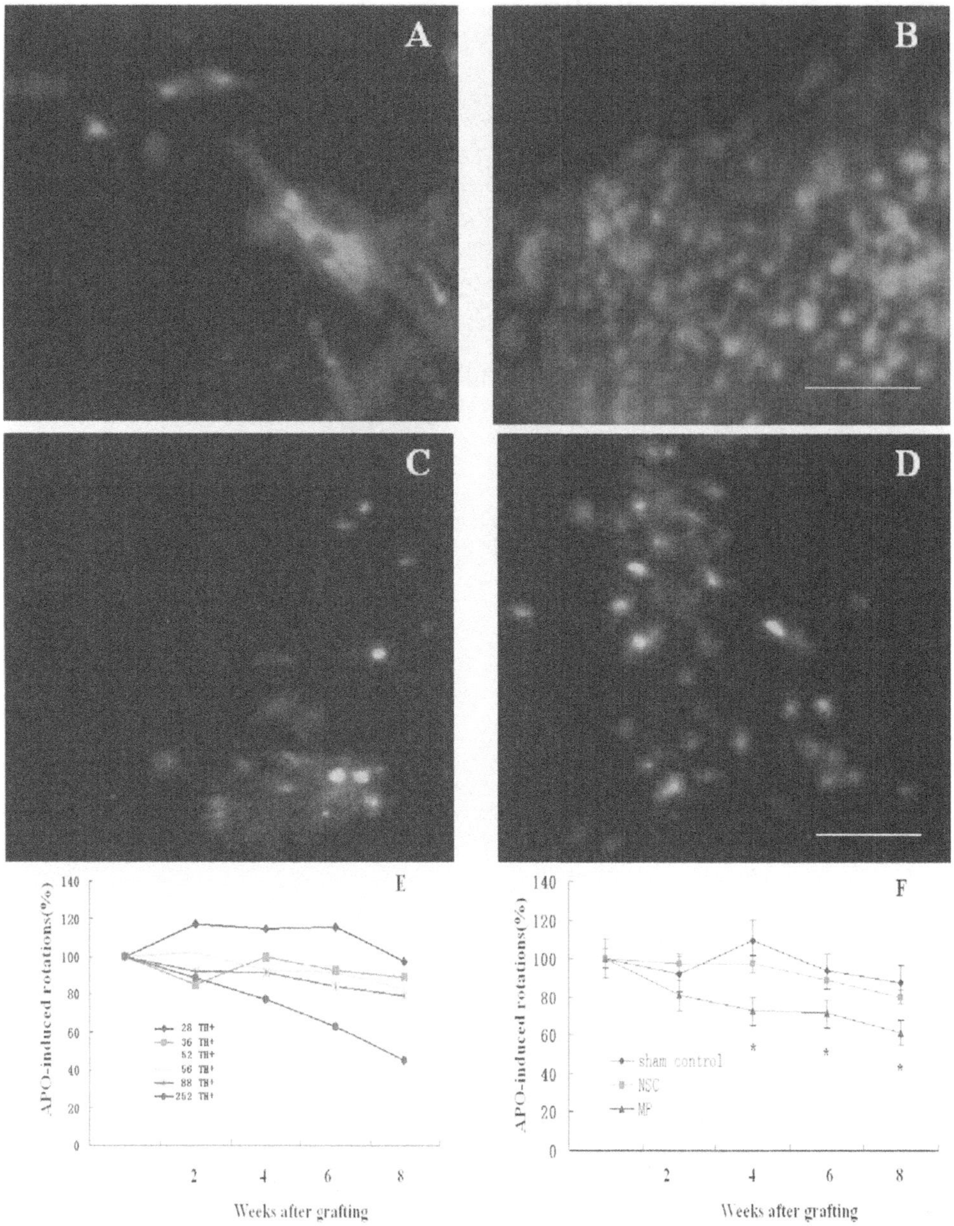

Fig. 2. The survival and differentiation of NSCs/MP grafted into intact and 6-OHDA-lesioned striatums and their effects on PD rats. A–D showed DiI-labeled NSCs (A and B) and TH-positive neurons (C and D) in striatum. In intact striatums, few grafted NSCs survived (A) and TH-positive neurons differentiated from MP (C). But in striatums lesioned with 6-OHDA, the livability and differentiation increased greatly (B and D). (Scale bar: 100 μm). E and F showed the trend of rotation behavior along with the time and the number of TH-positive neurons in the transplantation sites followed by being implanted into NSDs/MP. Figure E led to the conclusion that improvement degree of rotation behavior concorded with the number of TH-positive neurons detected in transplantation sites: the more TH-positive neurons, the greater relief of rotation behavior. F shows the different therapeutic effects of donor cells. PD rats grafted with NSCs showed little difference from control animals. But PD rats grafted with MP greatly ameliorated their rotation behavior 4–8 weeks later. (The asterisk signifies $P < 0.05$ of MP transplantation compared to control group and NSCs group)

planted into the striatum 8 weeks later with the rotation behavior continuously examined for 8 weeks. It had been found that NSCs had no ameliorating effect on the rotation behavior ($P > 0.05$) but the rotation behavior of one PD rat had decreased by more than 50%. The improvement of rotation behavior in PD rats receiving NSCs/MP transplantation was related to the number of differentiated TH-positive neurons. That is, the more the differentiated TH-positive neurons, the greater were the improving effects (Fig. 2E). Finally we compared the effects of NSCs and MP transplantation on PD rats. It was found that MP had more potent effects than NSCs. PD rats receiving MP transplantation displayed significant behavioral improvement after 4 weeks as compared with the sham controls or with the NSCs transplantation (Fig. 2F). The effects persisted for at least 8 weeks.

Discussion

Since NSCs/MP may be used potentially in cell transplantation for PD [6, 7], the survival, differentiation and maturation of NSCs/MP in vitro or in intracerebral environment may be mainly concerned. The data presented here demonstrated that NSCs/MP could proliferate freely in vitro and possessed the capacity to differentiate into neurons, such as DAergic neurons. An interesting result of this study is the significant difference of NSCs/MP survival and differentiation after being grafted into normal and PD rat striatums. It appears that intact striatum could not provide the proper environment for NSCs. In that environment, TH-positive neurons were very few. On the one hand, a reason may be that few grafted cells survived up to eight weeks. On the other hand, it may be due to the difficulty for the surviving grafted cells to differentiate into DA neurons. However, after being grafted into 6-OHDA-lesioned striatum, the ratio of the vital and differentiated cells was much higher four weeks later. Consistently, NSCs differentiated into much more TH-immunopositive neurons in PD groups than in control groups. Obviously, a lesioned striatum is more favorable for survival and differentiation of grafted cells than an intact one. It has been reported [4] that once the nigrostriatal dopaminergic pathway is destroyed, intracerebral microenvironment would change in lesioned striatum, such as synthesis of neurotrophic factor [4, 5], other neuropeptides [12] and glia proliferation. Protein extracts from the corresponding sites have potent neurotrophic effects and increase the number of DAergic neurons in vitro, which is ascribed to their role in increasing the survival of cultured cells and directly inducing non-DAergic neurons to DAergic neurons [4]. In the present study, it was observed that after NSCs were grafted into the striatum of PD model rats, their survival and differentiation were improved so as to greatly increase the number of the TH-immunopositive neurons. Perhaps the results are related to the mentioned factors. Above all, it shows more promising prospect to treat neurodegenerative diseases using grafted NSCs induced with special purified factors. The mechanisms of the factors remain to be determined.

In this study, NSCs grafted into PD rats can differentiate into TH-immunopositive neurons. But because the number is low, they can't markedly ameliorate behavior disorder. Among them, one PD rat appeared to recover visibly. With the immunohistochemistry method 252 TH-immunopositive neurons were detected in its transplantation site. They all could produce and secret DA in the transplantation sites. In the other 5 PD models, only 28–88 TH-immunopositive neurons were detected. So the rotation behavior decreased by less than 20.91%. It is concluded that behavior amelioration is consistent with secreted DA. Comparing the PD groups grafted with NSCs/MP with the control group, we found that different donor cells led to different rotation behavior recovery. The animals grafted with NSCs showed few differences from the control ones. But the rats grafted with MP showed obvious rotation behavior improvement and this effect lasted up to the 8th week. These results were consistent with what was observed in vitro: the minority of NSC differentiated into TH-positive neurons, but MP differentiated into TH-positive neurons at a higher rate cultured with serum-contained medium. This is mainly because MP contains primary DA neurons derived from SN and ventral tegmental area. They hold the potential to differentiate into TH-positive neurons even after being cultured in vitro. And 10 days after being propagated in vitro, some primary DAergic cells are alive in the clones. So more MP differentiated into TH-positive neurons and improved the rotation behavior more distinctly. As we know, NSCs are good candidates for multiple cell-based therapies for neural diseases by means of their biological properties. In short, it can be concluded that NSCs prefer to be induced into MP using appropriate factors before transplantation. Hence it is not a better way to treat PD only with NSCs. A combination of induction in vitro and MP transplantation should be considered more attentively in future studies. Appropriate and sufficient dopaminergic neurons and other neural progenitor cells from neural stem cells via cell induction in vitro could be provided for treating PD and other various neurodegenerative disorders in clinical application [10, 11].

Additionally, rotation behavior began to be ameliorate 2–4 weeks after being grafted with NSCs/MP. The latency is longer as opposed to what had been reported on primary DAergic neurons to improve rotation behavior just 1 week after being grafted. It is likely that more time is needed for NSCs to propagate in vitro and adapt to the transplantation sites before coming into play. On the other hand, NSCs/MP are in the mitotic stage and special conditions are required to divert to differentiate and mature into divided primary cells. So, if the NSCs develop to a subsequent stage after mitosis in vitro before transplantation, the latency period is supposed to be shortened in the treatment of neurodegenerative diseases.

References

1. Bjorklund A, Lindvall O (2000) Cell replacement therapies for central nervous system disorders. Nat Neurosci 3(6): 537–544 (review)
2. Gage FH, Coates PW, Pallmer TD *et al* (1995) Survival and differentiation of adult neuronal progenitor cells transplanted to the adult brain. Proc Natl Acad Sci USA 92: 879–883
3. Gan WB, Grutzendler J, Wong WT *et al* (2000) Multicolor "DiOlistic" Labeling of the Nervous System Using Lipophilic Dye Combinations. Neuron 27(2): 219–225
4. Hida H, Fukuda A, Fujimoto I *et al* (1997) Dopamine-denervation enhances the trophic activity in striatum: evaluation by morphological and eletrophysiological development in PC12D cells. Neurosci Res 28: 299–221
5. Hyman C, Juhsz M, Jackson C *et al* (1994) Overlapping and distinct actions of the neurotrophins BDNF, NT-3, and NT-4/5 on cultured dopaminergic and GABAergic neurons of the ventral mesencephalon. J Neurosci 14: 335–347
6. Isacson O, Costantini L, Schumacher JM *et al* (2001) Cell implantation therapies for Parkinson's disease using neural stem, transgenic or xenogeneic donor cells. Parkinsonism Relat Disord 7: 205–212
7. Lindvall O (2001) Parkinson disease. Stem cell transplantation. Lancet [Suppl] 358: S48
8. Price J, Williams BP (2001) Neural stem cells. Curr Opin Neurobiol 11(5): 564–567
9. Peschanski M (2001) 10 years of substitution therapy for neurodegenerative diseases using fetal neuron grafts: a positive outcome but with questions for the future. J Soc Biol 195(1): 51–55. French
10. Sanchez PR, Studer L, Bankiewicz KS *et al* (2001) In vitro generation and transplantation of precursor-derived human dopamine neurons. J Neurosci Res 65(4): 284–288
11. Svendsen CN, Rosser AE (1995) Neurons from stem cells? Trends Neurosci 8: 24–34
12. Xie JX, Sun ZHH, Jiang ZHY *et al* (2000) Effects of cholecystokinin octapeptide and lesion of the medial forebrain bundle on the contents of dopamine and CCK-8 from ventral tegmental area and nucleus accumbens. Chin J Neurosci 16(1): 13–16

Correspondence: Zuoping Xie, Institute of Neuroscience of Department of Biological Sciences and Biotechnology, Tsinghua University, Beijing 100084, China. E-mail: zuopingx@mail.tsinghua.edu.cn

Acta Neurochir (2003) [Suppl] 87: 175–180

Mesencephalic progenitors can improve rotational behavior and reconstruct nigrostriatal pathway in PD rats

Z. H Sun[1], Y. L Lai[1], W. W. Zeng[1], D. Zhao[1], Z. W. Ye[2], H. C. Zuo[2], and Z. P. Xie[1]

[1] Department of Biological Sciences and Biotechnology, Tsinghua University, Beijing, China
[2] Medical School, Tsinghua University, Beijing, China

Summary

The aim of this study was to investigate the possibility of mesencephalic progenitors (MP) in treating Parkinson's disease (PD). MP were prepared from E_{11-13} rats and proliferated in serum-free medium with basic fibroblast growth factor (bFGF) for 10 days. Cells were then collected and implanted into the striatum only – single grafts or simultaneously into the substantia nigra (SN) and the striatum – double grafts. Twelve weeks after transplantation, DiI, a fluorescent dye, was microinjected into the ipsilateral striatum. Using this strategy, it was found that MP of double grafts had more potent effects on rotational behavior than that of single grafts. Injection of the retrograde tracer DiI into the striatum resulted in fluorescent-labeled cells within the intranigral grafts in double grafts. These data greatly support that MP transplants could not only improve rotational behavior, but does help to re-establish nigrostriatal connections so that it may become one efficient way in treating PD.

Keywords: Mesencephalic progenitor cells; brain grafts; Parkinson's disease; tyrosine hydroxylase.

Introduction

At present there are several strategies to be used in treating Parkinson's disease (PD), such as drug therapy, cell therapy, and gene therapy. Among them, cell therapy means more to us. It is potentially the most promising way to all patients [1]. Cell transplantation using tissues from the fetal ventral mesencephalon (VM) offers a possible therapeutic approach to treat PD [7, 9]. However, widespread application is severely limited by the practical problems of obtaining sufficient donor tissues from human embryos. One alternative source of fetal donor cells for clinical cell therapy of PD is basic fibroblast growth factor (bFGF)-responsive mesencephalic progenitors (MP). MP from embryonic brain of normal Sprague-Dawley (S.D) rats have self-renewable and multipotent properties and thus are good candidates for donor cells in neural transplantation [2, 8].

The traditional methods in neural transplantation for PD are single grafts placed in the striatum or other brain areas instead of substantia nigra (SN). Ectopically placed embryonic grafts can extend axons, form synapses in the host striatum, release dopamine (DA) and improve dysfunction in PD, but they can't reestablish normal nigrostriatal pathway [10]. Furthermore, ectopically placed grafts fail to rebuild dendritic DAergic modulation to the SN in the pars reticulate (SNr), which is important in the regulation of basal ganglia outflow. The lack of host-graft connectivity and dendritic release of dopamine are likely to be responsible for limiting complete symptomatic recovery in transplant recipients. Mendez *et al.* [4] reported the reconstruction of the nigrastriatal pathway by simultaneous intrastriatal and intranigral DAergic transplants. They found that double grafts of primary DAergic neurons promoted the growth and guidance of axons from a similar graft placed ipsilaterally in the SN. Since primary neurons are capable of restoring nigrastriatal pathway, bFGF-responsive MP which are able to differentiate into DAergic neurons might exert the same effects. In our study, we attempted to implant MP into SN and striatum simultaneously in PD rats so as to find a suitable substitute for primary DAergic neurons to treat PD, which would also play a key role in reconstructing nigrostriatal pathway.

Materials and methods

Animals and dissections

Timed pregnant Sprague-Dawley rats were used for cell cultures or cell grafts. Insemination was confirmed by vaginal plug and considered as E_0. Rats (n = 18) had a unilateral 6-OHDA lesion in the nigrostriatal pathway. Among them, six (n = 6) received intra-CPu transplants (single grafts), six (n = 6) received simultaneous intranigral and intrastriatal transplants (double grafts). The other rats (n = 6) were assigned to control groups, which received medium injection only.

Cell preparation for cultures and grafts

Cells were dissociated from E_{11-13} rats ventral mesencephalon. Cells were plated at 5×10^5/25 mm flask in serum-free DMEM + N_2 (GIBco) medium containing 20 ng/ml bFGF (Sigma). Ten days later, cells were collected and primed with dbcAMP (Sigma) for 1 day to differentiate. Before transplantation, cells were dissociated mechanically until a homogenous single cell suspension was achieved. 4×10^5 cells were used with viability >95% determined by the trypan blue dye exclusion method.

Animal treatments and behavioral tests

180–200 g female S.D rats were injected with 20 μg 6-OHDA (Sigma) in the right CPu (AP = 1.8 mm and −0.2 mm; ML = −2.5 mm and −3.5 mm; DV = −6.5 mm and −6.0 mm according to Paxinos and Watson atlas [6]). Five weeks after treatment, the level of nigrostriatal pathway lesion was tested by measuring the motor activity of treated animals after administration of 0.1 mg/kg apomorphine (APO, Sigma) [5]. The individuals that showed ipsilateral rotation score > 7 were selected for study. Eight weeks after lesion, 4×10^5 MP cells were injected into the 6-OHDA-lesioned striatum using a microsyringe or/and into SN at AP = −5.3 mm; ML = −2.5 mm; DV = −7.8 mm. At various intervals (2, 4, 6, 8, 12 weeks) after unilateral grafting, animals were injected with 0.1 mg/kg apomorphine and turning behavior was examined to evaluate DA functional recovery.

Immunocytochemistry

Twelve weeks after transplantation, rats were anaesthetized with sodium pentobarbital and sacrificed by transcardiac perfusion with 4% paraformaldehyde in 0.1 M PBS PH7.4, the brains removed, post-fixed and sectioned at the cryostat. Tyrosine hydroxylase (TH)-like immunoreactivity was detected with a mouse anti-TH antibody (1 : 10000, Sigma) and revealed by avidin-biotin method.

Nestin immunoreactivity was detected with a mouse anti-nestin monoclonal antibody (1 : 1000, Chemicon) and revealed by fluorescein(FITC)-conjugated goat anti-mouse IgG.

DiI injections [3]

At twelve weeks after transplantation, three animals with double grafts (n = 3) three animals with intrastriatal grafts (n = 3), three animals with 6-OHDA lesions only (n = 3) and three controls (no lesion or grafts) (n = 3) received stereotactic injections of the retrograde neuronal tracer 1,1′-Dioctadecyl-3,3,3′,3′-tetramethylindocarbocyanine perchlorate (DiI, 2.5 mg/ml in 100% ethanol). DiI was injected into the striatum (1 μl per injection) using a microsyringe at the above coordinates. Two weeks after the injections, animals were anesthetized, perfused, and brains sectioned in the same manner as described for immunohistochemistry. One-fourth coronal sections were mounted and analyzed for DiI (Biotium) labeling using a fluorescence microscope equipped with filters rodamine for 546 nm excitation. Adjacent sections were stained for TH as described above.

Results

Mesencephalic cell clone expansion

After being cultured in serum-free medium supplement with bFGF, MP proliferated well, making floating neurospheres/clones that were nestin-immunopositive (Fig. 1A and B). After passage of one generation (5–7 days in vitro), the numbers of total cells propagated more than 10 times. Once the clones stuck to the P-D-lysine treated plate and primed with dbcAMP following withdrawal of bFGF, the cells began to differentiate into neurons and glia, many of which were TH-immunopositive neurons (Fig. 1C). The cell proliferation could be maintained at least for 2 months.

Examination of TH-immunoreactivity neurons after transplantation

Three months after transplantation, immunostaining revealed that the grafted MP survived and differentiated into TH-immunopositive neurons which appeared in transplantation sites (Fig. 1D). They did not distribute diffusely or homogeneously in the grafts. They made aggregation formation around the transplantation track. Similar numbers and phenotype of TH-immunopositive neurons were observed in the striatum both in single and in double grafts.

Intrastriatal DiI injection

Injection of DiI into the striatum resulted in diffuse labeling throughout the injection sites including the intrastriatal grafts. Numerous retrograde labeled cells were displayed in the ipsilateral SN two weeks later in animals bearing unilateral lesions and receiving simultaneous transplants. The fluorescent labeled cells were randomly distributed in the main body of the SN transplants. About 5.0% ± 1.1% of the TH-immunopositive cells were also found to be fluorescent (Fig. 2A and B). The remaining neurons which were immunoreactive for TH were not fluorescent cells. Also, a portion of the cells (67.5% ± 2.7%) were fluorescent labeled but not immunoreactive for TH. No labeling was noted in the sham control animals which

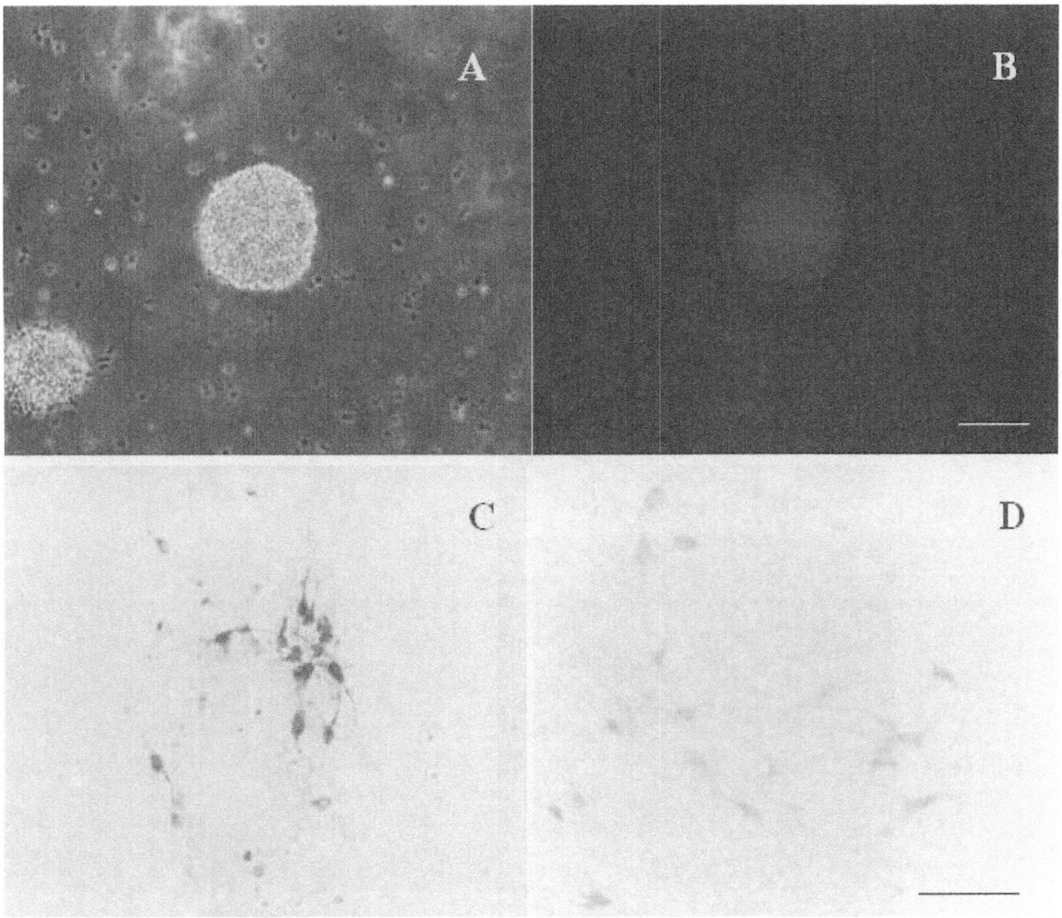

Fig. 1. Mesencephalic progenitor neurospheres exhibited nestin-positive (A and B). A was the phase contrast image of bFGF-responsive mesencephalic progenitor neurospheres. B was the fluorescent image of nestin-positive MP. (Scale bar is 100 μm). Mesencephalic progenitors can differentiate into TH-immunopositive neurons (C) after serum-free 2 days culture in vitro while dbcAMP was added or (D) in transplantation site after 12 weeks post transplantation. (Scale bar is 100 μm)

were lesioned only and in the animals which had received single grafts in the striatum at the corresponding sites (Fig. 2C and D).

Behavioral tests

Most animals bearing unilateral 6-OHDA lesions of the nigrostriatal pathway demonstrated asymmetric turning behavior in response to APO (0.1 mg/kg, s.c.) challenge. At two weeks post-transplant, APO-induced asymmetric rotational behavior was almost no reduction. They were 19.19% ± 7.11% (single grafts) and 25.35% ± 7.64% (double grafts) reduction compared to that in pre-transplantation, respectively. At eight weeks post-transplant behavior, recovery was still at moderate decrease: 38.63% ± 8.07% (single grafts) and 39.51% ± 3.43% (double grafts). The difference did not reach significance until the MP had been implanted for about 12 weeks. The sham control animals continued to exhibit contralateral rotations when injected APO (Fig. 2E).

Discussions

Progenitor cells can proliferate freely in serum-free N_2 medium with bFGF. After they were grafted into the brain of fetal or neonatal animals, they migrated widely in the brain and differentiated into different phenotypic neurons, matching their target areas. MP can be characterized by a kind of progenitor cells from embryonic mesencephalon which includes A9 (substantia nigra) and A10 (ventral tegmentental area) DAergic neurons [2, 7, 8]. In the present study it was found that MP have similar features to all types of progenitor cells (self-renewable) but tend to differentiate into dopaminergic neurons in vitro or in vivo.

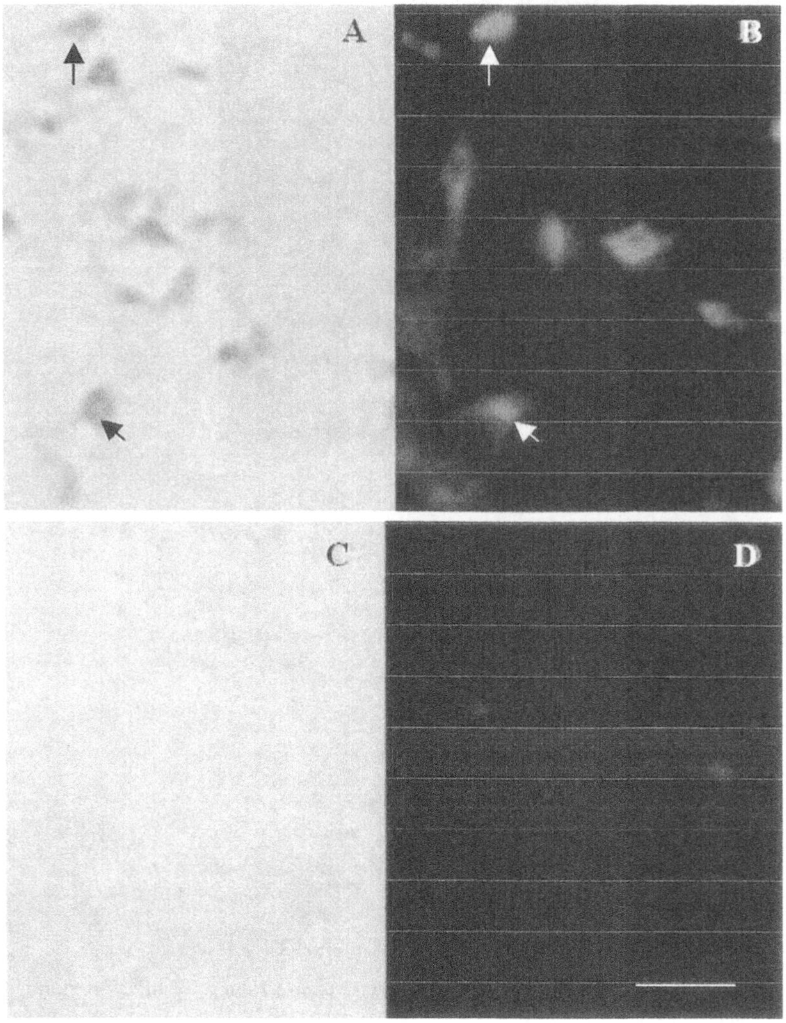

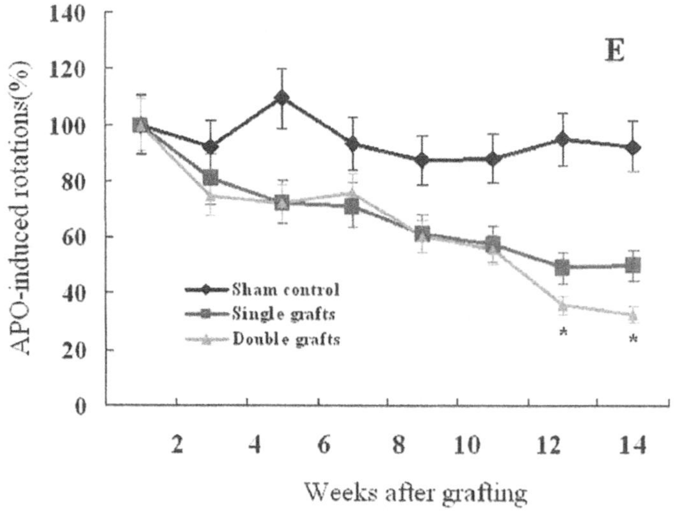

Fig. 2. TH-immunopositive neurons (A and C) and DiI-labeled neurons (B and D) in substantia nigra. They were observed in double grafts after the retrograde tracer had been injected into the ipsilateral striatum for 2 weeks following 12 weeks post MP transplantation. The arrows in (A) and (B) indicate that TH-immunopositive neurons' axons generated from the transplanted MP in SN reached the striatum so that DiI-labeled somites were observed in SN. There were no TH-immunopositive and DiI-labeled neurons in the single-graft groups (C and D). In the sham control group, similar results appeared as in the single-graft group (Scale bar is 50 μm.). (E) represents the respective effects of single grafts and double grafts of mesencephalic progenitors on rotational behavior in PD rats. Double grafts have more potent effects on rotational behavior in PD rats than single grafts after MP has been implanted for 12 weeks. There was no significant difference between the two groups from 2–8 weeks' transplantation. The percentage used of the APO-induced rotations in post-graft animals was relative to the total number of that in pre-graft animals. Asterisk, $P < 0.05$ compared to single graft group

They could be expanded under bFGF for at least 2 weeks before transplantation and they were nestin-immunopositive (>85%) – a marker for progenitors. This suggests that most of the cells being propagated were neural progenitor cells, although the probability that among these progenitors there were some primary DAergic neurons is not excluded completely. Not only were the proliferated cells able to differentiate into TH-immunopositive neurons – a marker for DAergic neurons – treated with 10% fetal bovine serum and primed by dbcAMP for 1 or 2 days in vitro, but they could differentiate into mature TH-immunopositive neurons and ameliorate APO-induced rotation after being transplanted into 6-OHDA-lesioned sites of PD rats. These data suggest that MP prepared in our lab are credible and can be used for further investigation.

It is known that the degeneration of the SN in PD results in a loss of dopaminergic innervation to the striatum. Although transplanted DA cells placed in the SN can release DA *in situ*, they cannot grow axons over long distances to reach the ipsilateral striatum, nor synapse with the corresponding neurons and function as the physiological neurotransmitter. So intrastriatal grafts which can lead to DAergic reinnervation of the host striatum will substitute for intranigral ones in clinical application. The intrastriatal grafts can improve simple movement function, such as asymmetric rotational behavior in the 6-OHDA-lesioned PD rats, but they cannot completely restore the more complex sensorimotor behavioral defection because they cannot receive SNr input and thus lose the dendritic connection to the adjacent structure in SN. Mendez *et al.* [4] demonstrated for the first time that simultaneous DAergic transplants placed in the substantia nigra and ipsilateral striatum-double grafts-of PD rats could reconstruct the DAergic nigrostriatal pathway similar to the physiological conditions. In this study, we attempted to implant bFGF-responsive MP instead of primary DAergic neurons and investigated if they could reconstruct the nigrostriatal pathway by simultaneous intranigral and intrastriatal grafts. Just as we had expected, the MP were similar to embryonic primary VM transplants. Their axon outgrowth arising from a graft homotopically placed in the substantia nigra was capable of extending to the ipsilateral striatum. It may be ascribed to the neurotropic effects of MP placed in the ipsilateral striatum, which can direct the axon growth along the physiological pathway. Additionally, stronger behavioral improvement was also observed in animals with double grafts of MP than in animals receiving a single intrastriatal graft. In conclusion, the MP placed in the striatum not only release dopamine to reinnervate the postsynaptic neurons, but probably simulate the natural environment in the target to induct the axon growth of MP placed in the ipsilateral SN as a neurotropic factor. It has been found that in animals which received double grafts, only about 5.0% ± 1.1% of the TH-immunopositive cells appeared to be fluorescent. This may be due to the fact that the other TH-immunopositive neurons' axons did not reach the striatum, so no corresponding DiI-labeled somites were found. Otherwise, the rest DiI-labeled neurons may be of other cell types, such as acetylcolinergic, GABAergic or glutaminergic neurons etc, so they were not TH-immunoreactive.

Mendez *et al.* [4] found that amphetamine-induced asymmetric rotation recovery was apparent at 6 weeks post-transplant in models of double grafts. Unlike their results, the present study demonstrated that double grafts did not have more potent effects on rotational behavior than single grafts in PD rats until the MP had been implanted for 12 weeks. The differences are supposed to result from the different rotation inducers – the former was APO-induced (a drug as DA receptor agonist) and the latter was amphetamine-induced (a drug as catecholamine release inducer) rotational behavior. Another important factor was that the donor cells in our study were bFGF-responsive MP instead of primary DAergic neurons. They are mitotic neurons but not post-mitotic. Thus MP needed a little longer to adapt to the intracerebral environment and to start differentiating than primary DAergic neurons.

In conclusion, MP proliferated and differentiated into TH-immunopositive neurons in vitro and single/double grafts in vivo will be useful for further studies aimed at treating PD. In addition to MP, various neural stem/progenitor cells are expected to substitute for primary DAergic neurons after cell culture and induction by specific factors in vitro. The therapeutic effects of neural stem/progenitor cells after being induced to TH-immunopositive neurons should be substantiated depending on this system.

References

1. Bjorklund A (2000) Cell replacement strategies for neurodegenerative disorders. Novartis Found Symp 231: 7–15; discussion 16–20 (review)
2. Carvey PM, Ling ZD, Sortwell CEA (2001) Clonal line of mesencephalic progenitor cells converted to dopamine neurons by

hematopoietic cytokines: a source of cells for transplantation in Parkinson's disease. Exp Neurol 172(1): 98–108
3. Honig MG, Hume RI (1986) Fluorescent carbocyanine dyes allow living neurons of identified origin to be studied in long-term cultures. J Cell Biology 103: 171–187
4. Mendez I, Sadi D, Hong M (1996) Reconstruction of nigrostriatal pathway by simultaneous intrastriatal and intranigral DAergic transplants. J Neurosci 16(22): 7216–7227
5. Moore AE, Cicchtti F, Hennen J, Isacson O (2001) Parkinsonian motor deficits are reflected by proportional A_9/A_{10} dopamine neuron degeneration in the rat. Exp Neurol 172(2): 363–376
6. Paxinos G, Watson C (1982) The rat brain in stereotaxic coordinates. Academic Press, London
7. Ptak LR, Hart KR, Lin D, Carvey PM (1995) Isolation and manipulation of rostral mesencephalic tegmental progenitor cells from rat. Cell Transplant 4(3): 335–342
8. Rina SS, Jauniaux E, Stern GM, Bradford HF (2002) The controlled conversion of human neural progenitor cells derived from foetal ventral mesencephalon into dopaminergic neurons in vitro. Brain Res DevBrain Res 30: 136(1); 27–34
9. Son JH, Chun HS, Joh TH, Cho S, Conti B, Lee JW (1999) Neuroprotection and neuronal differentiation studies using substantia nigra dopaminergic cells derived from transgenic mouse embryos. J Neurosci 19(1): 10–20
10. Zhou FC, Chiang YH, Wang Y (1996) Constructing a new nigrostriatal pathway in the parkinsonian model with bridged neural transplantation in substantia nigra. J Neurosci 16: 6965–6974

Correspondence: Zuoping Xie, Department of Biological Sciences and Biotechnology, Tsinghua University, Beijing 100084, China. E-mail: zuopingx@mail.tsinghua.edu.cn

Author Index

Index of Keywords

Springer-Verlag
and the Environment

GPSR Compliance

The European Union's (EU) General Product Safety Regulation (GPSR) is a set of rules that requires consumer products to be safe and our obligations to ensure this.

If you have any concerns about our products, you can contact us on ProductSafety@springernature.com

In case Publisher is established outside the EU, the EU authorized representative is:

Springer Nature Customer Service Center GmbH
Europaplatz 3
69115 Heidelberg, Germany

Batch number: 10731490

Printed by Printforce, the Netherlands